weather wise
reading weather signs
alan watts

D1354742

ADLARD COLES NAUTICAL • LONDON

Published by Adlard Coles Nautical
An imprint of A & C Black (Publishers) Ltd
38 Soho Square, London W1D 3HB
www.adlardcoles.com

First edition 2008

ISBN: 978-0-7136-8153-6

A CIP catalogue record for this book is available from the
British Library.

This book is produced using paper that is made from wood
grown in managed, sustainable forests. It is natural, renewable
and recyclable. The logging and manufacturing processes
conform to the environmental regulations of the country of origin.

Designed by James Watson
Typeset in 9/11.5pt Helvetica Light

Printed and bound in Spain by GraphyCems

Note: while all reasonable care has been taken in the publication of
this edition, the publisher takes no responsibility for the use of the
methods or products described in the book.

contents

about this book

This is a weather book with a difference. It sets out to answer, in simple terms, the questions that so many people ask about the weather. Why is the sky blue? Why does it rain? What happens in the clouds and why don't all clouds rain? Why does the wind get up with the day and die down at night? How do I tell if bad weather is coming? How do I reduce the risk of being struck by lightning? What should I beware of when I leave the lowlands and walk or climb higher?

I have liberally illustrated these chapters with photographs from the collection I have made over the years. They have been chosen not only because they illustrate aspects of the weather but also because they are, in many cases, photogenic. The sky and weather phenomena generally are wonderful subjects for the photographer,

but the sky changes from moment to moment, and having your camera with you at all times will enable you to capture truly beautiful skies or fascinating phenomena.

After a lifetime spent trying to explain weather to people who are not, and never wish to be, meteorologists, I have had many questions fired at me. In this book I have tried to answer some of these questions as well as many others.

A previous book of mine, *Instant Weather Forecasting*, is a kind of ready-reckoner for weather conditions that will usually accompany the 24 specially selected sky pictures. I wrote it to suggest what each characteristic sky might mean for the coming hours. *Weather Wise* is different. It gives more explanations and provides examples of the weather experienced in the temperate latitudes of the northern hemisphere.

For all those who live in climes to which this book applies, I hope you will find answers to what you want to know between these pages. If you want something more in depth, then try *The Weather Handbook* which I wrote to give the interested layman an insight into how the weather behaves.

Alan Watts

1 the seasons

The seasons follow the sun so we will start where many ancient religions started their year – at the winter solstice. By the calendar, this is on or about 22 December. Ancient priests observing sunrise would, during December, have seen sunrise creep slowly further and further southwards day by day until it came to a standstill for several days around 22 December. On or around 25 December there would be a glimmer of movement back towards the north again. The Celtic tradition held that the sun was reborn at this time.

Gradually the sun's risings move back further and further northwards until it is rising due east at the spring equinox (21 March). However, the weather of winter and early spring lags behind the sun. The sun may get higher but the winds remain cold and the days often get colder. For the first months of the year: 'As the days lengthen the cold strengthens'.

Photo 1 The typical sky of spring – big cumulus clouds, showers and blossom. The air is often cool and unstable.

Photo 2 What we would all like in summer: fair-weather skies, balmy winds and delightful surroundings.

The sea, which has been getting colder since September, continues to do so, and this often governs the temperature of the air, so that January and February are usually the coldest months. In the worst years, March is not much better – it may even be colder than the previous months – and spring is slow to start. Yet it is a very unusual year if spring is not bursting forth by the spring equinox.

Late spring and early summer see the sunrise moving northwards day by day but slowing down as the summer solstice approaches. Sunrise observers see the sun come up more and more closely to the furthest northerly point it will achieve. Its sunrise position then stands still for several days around 21 June before it slowly begins to move back south through the autumnal equinox until it again stands still at the next winter solstice. The summer solstice is often referred to as midsummer but in the northern hemisphere we know that the best summer weather is yet to come. This yearly rhythm of the sun controls the amount of heat a hemisphere acquires, but it is the sea that absorbs that heat in depth, and this means that it is late August or early September before the sea is at its warmest. The sea is usually coldest in March and early boaters should beware.

In the temperate latitudes cold winds from northern seas and polar regions keep spring cold, showery and often blustery. Equally it is the warmest sea temperatures of the year that lead to autumn being the 'season of mists and mellow fruitfulness'. Summer is when the bad weather systems migrate northwards leaving central and southern temperate latitudes warm and usually with light winds. Not every time, though.

It was high summer, August 1979, when, in light winds, a great fleet of yachts left the Solent in southern England to take part in the biennial Fastnet Yacht Race. The race is so-named because the turning mark is the Fastnet Rock off southern Ireland and the course carries the fleet into very exposed waters. So when an unprecedented vicious storm blew up, with most yachts strung out between the Scilly Isles and Fastnet, what should have been a happy competitive event turned into a disaster that claimed the lives of 15 crew and led to 20 yachts being abandoned.

There are sometimes strong winds in this area in August, but only very occasionally do storms more reminiscent of January come along. It was 17 years (1956) since a similar kind of storm in the English Channel had prompted the Royal Ocean Racing Club to invoke an enquiry into weather that decimated an ocean race. When you realise how this area, the centre of English yachting, is alive with sails through the summer, it is evident that big storms are very rare here.

Yet they do occur and there is no way of knowing if another might turn up. The changeable weather of the latitudes surrounding 50°N never allows people to be complacent. January may be benign and almost warm, while it snows (on rare occasions) in June. Luckily there are very good weather forecasting services these days and people who are not forewarned have only themselves to blame.

Why are there seasons?

Seasons are easier to understand if you examine a simple model. Take a dinner plate whose rim represents the path the Earth takes as it orbits the sun. Raise it a bit – a thick book is useful – and get a ball (eg an orange) to represent the Earth. Stick a knitting needle through the ball to represent the axis about which the Earth spins (Fig 1.1). Draw a line round the ball to represent the Equator.

Incline the Earth's axis as in Fig 1.1 and move it round the rim, making sure you keep the axis pointing in the same direction in space. This is crucial because the Earth is like a gyroscope and always maintains its direction of spin (Fig 1.2).

Midwinter in the northern hemisphere (NH) is at A. The sun's rays strike the NH obliquely and so less heat is absorbed by the hemisphere and it is coldest. Conversely it is midsummer in the southern hemisphere (SH) and the sun is higher in the sky. Move round a quarter of the orbit and it is NH spring. Now all the places on the same latitude see the sun equally. It is the spring equinox and everywhere the sun rises due east and sets due west. It does so at 6 am and 6 pm local solar time (time measured from midday when the sun crosses the meridian).

In another three months the sun is highest in the sky in the NH and it is the summer solstice. For example at Stonehenge, where the edifice is orientated to the summer solstice, the sun rises about 40° north of east and sets the same number of degrees north of west. In the SH it is the winter solstice, the name stemming from the sun's rising position standing for several days in its most northerly position.

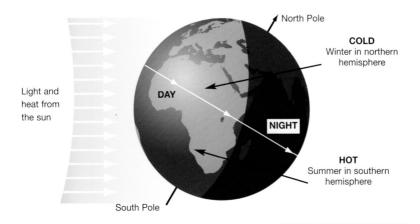

Fig 1.1 Because the Earth's axis is tilted, the energy from the sun strikes the two hemispheres differently through the year. The situation shown is when it is winter in the northern hemisphere and summer in the southern. We see why it is continually night for months in the regions around the North Pole and why, at this time of year, there is continal daylight at the South Pole. In six months the situations are completely reversed.

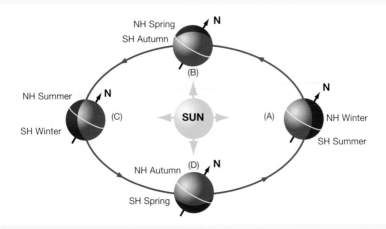

Fig 1.2 As the Earth moves round its orbit, the situation in Fig 1.1 is (A). Those in the northern hemisphere are at the winter solstice. Three months later (B), both hemispheres see the sun equally and the length of the day is the same as that of the night all over the world. It is the spring equinox in the northern hemisphere. A further three months brings us to (C) and it is the summer solstice in the northern hemisphere and the winter solstice in the southern. Three months later (D), again the two hemispheres see the sun equally – it is the autumnal equinox in the northern hemisphere. Thus we get the progression of the seasons in both hemispheres.

Photo 3 In autumn the late afternoons tend to be calm and low mists form under cooling skies.

Then around 21 September the Earth has moved on another quarter of its orbit and again we are at an equinox – the autumnal equinox. The year is completed when once again the festivities to mark the newborn sun follow the winter solstice.

The rhythm of land and sea

The land surface absorbs the heat of the sun and so gains least around 21 December. Then through the spring it gains more heat energy and some of this is transported down into the soil, which continues after the summer solstice. However, it is autumn, visited by cooler winds, when the storage heater of the Earth gives us back its stored heat and keeps us warmer than we might expect and for much longer.

It has always been like that – for as Keats says in his *Ode to Autumn*, 'Until they think warm days will never cease' – warmth hangs on much longer than you'd expect.

The sea is a much more efficient storage heater than the land but it gains its heat in depth, so its change in temperature is slow and it is September before the resorts see any real dip in the temperature of beach waters.

It is spring and early summer that sees the greatest contrast between coastal land and sea temperatures and it is the latter that drives seabreezes. (Seabreeze here specifically means the wind generated by the coast, as opposed to a sea breeze, which can be any wind that happens to be blowing from sea to land.) Seabreezes are coastal winds that blow from sea to land mainly between April and September. However, by September the difference between coastal land and water temperature is least, and it needs the most advantageous conditions before a breeze will blow. These conditions include light morning winds and lots of sunshine. In May and June, breezes can roll back opposing winds for 50 miles (80 km) but it would be very rare to find seabreeze winds 30 miles (50 km) inland in late August and September.

Photo 4 A late winter snowfall is unwelcome to spring blossom.

In many coastal places seabreezes ameliorate the climate, cooling the heat of summer in which people inland are sweltering. They also aid hay-fever sufferers by bringing in clean pollen-free air off the sea. However in spring they can produce snow showers as they move inland.

There are also seasonal winds that blow from the land to the sea and reach their greatest strength overnight. These land breezes are usually light but become substantial when high ground flanks the sea. The months when these are most likely are where the sea temperature is higher than the land ie in autumn and winter. However, overnight land breezes can blow at most times of the year.

If you live near the coast, you will often notice when you wake it is clear or partly cloudy but the ground shows that it has been raining. These early-morning coastal showers are usually due to cold air draining off the land overnight. The coastal sea is warmer than the air and it sets off showers, which clear when the sun is up.

Take a morning walk

It's winter and a fine sunny morning. Not a cloud in the sky and hardly any wind. It was chilly overnight – maybe a frost. You think: 'What a good day for a walk!' The advice is to go before lunch and not leave it to the afternoon. It often happens that by midday, or just after, heap clouds will have built and at the same time a chilly wind will blow up. What was a pleasant winter morning is quite forbidding by the afternoon. What has happened is that the sun has taken all morning to warm the ground sufficiently to produce convection. Both cloud and wind are the result.

2 a day's weather

Most days it gets warmest around the middle of the day and then the temperature drops towards the evening. We also notice that it is usual for the wind to increase, so that by the end of the morning there is a noticeable breeze even if it was close to calm at breakfast-time.

Photo 5 The dawn is red, the high sky is white. Bad-weather clouds gather, so do not be fooled by the gentle wind you have now; there is trouble on the way.

Well, that's the normal pattern except when the weather is 'changeable', and changeable weather comes most often when low-pressure systems cross the area.

In the early days of meteorology, lows or depressions were known as cyclones. This term is still used, but not often as it now specifically refers to the kind of hurricanes common in the Indian Ocean. However, we cannot avoid the term because-high pressure systems are still called anticyclones. When changes are taking place we describe those changes as being cyclonic, when the winds etc are going to change as they do when low-pressure systems are in charge. The changes will follow an anticyclonic pattern when high-pressure systems cover the area.

Lows and highs are described in chapter 6 but you can see how cyclonic winds behave by taking your left hand and rotating it anticlockwise. That's the way the winds move round centres of low pressure. Conversely, doing the same with your right hand shows how winds rotate about centres of high pressure, ie clockwise. (To remember which hand is which, highs usually bring the 'right' kind of weather.) If you live south of the equator, then these rules are reversed.

It will be the weather patterns that go with cyclonic and anticyclonic conditions that interfere with the normal way the weather elements change with the day. So let's see what happens on a normal kind of day. You can still expect a certain pattern of events to occur more or less on most days, whatever else is happening.

Right-now weather

Breakfast time is when most of us cast a weather eye out of the window or go into the garden to assess the prospects. Even if there is not a cloud in the sky, if it is cool and quite blustery you can expect showers later. This is very likely if the wind is from some point around north-west and there are already some ragged low clouds about.

The sun will set off convection and this will, in these conditions, produce heap clouds that grow into shower clouds during the day.

Frosty mornings under clear skies will normally stay fair for the day. However, beware increasing high cloud (see photo 14, page 40). It is possible that the coming night will be windy, cloudy and a lot warmer.

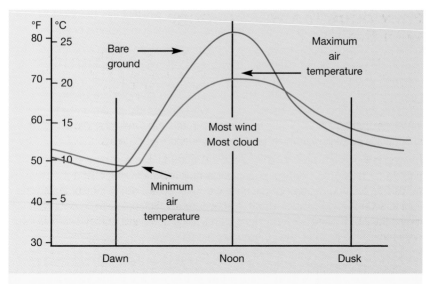

Fig 2.1 The way the ground and air temperatures change on a normal summer's day is the *diurnal variation in temperature*. Note that the air temperature is cooler than the ground temperature during the day and warmer than the ground by night. Ground temperature does what you would expect – it is maximum at local noon and minimum at dawn. The air temperature lags behind by an hour or two so that it is warmest in the early afternoon and coolest soon after dawn. The amount of cloud and the strength of the wind tend to follow the same pattern.

How does the temperature change during the day?

We can see this most clearly from Fig 2.1. There are two curves – blue for the ground and red for the air. The temperature over bare ground does what you'd expect; it is at its lowest at dawn and highest at local noon. It sinks away during the afternoon and through the night. However, the air reluctantly trails along behind, reaching its minimum temperature an hour or two after dawn and its maximum an hour or two after noon. So while the sun was highest in the sky when it was due south, the afternoon will get warmer for a couple of hours after this time.

In changeable weather, fronts will bring changes of air temperature as well as cloudiness, maybe interspersed with spells of sunshine, which will profoundly affect this simple pattern. In winter it may be warmer in the evening than in the middle of the day when a warm front passes during the afternoon.

How does the wind change with the day?

Briefly, wind speed follows the diurnal air temperature curve; it becomes lowest after dawn and rises during the morning to become strongest in the early afternoon, after which it sinks away into the evening. This is called the diurnal variation of wind speed.

While these diurnal changes are most easily appreciated when conditions are settled and there is very little wind overnight, it is noticeable that quite strong winds will blow most strongly in the afternoon and reduce during the night.

However, what happens depends on where you are. The diurnal changes shown are for land areas. Winds directed from an expanse of ocean will give the opposite effect – wind speed tends to be least in the afternoon and highest during the night. If you live on the coast facing an onshore wind, your diurnal variation will be that of the sea, not of the land.

Photo 6 Benign lumps of fair-weather cumulus cloud give a warm afternoon for swans and people.

If you live on the coast and have a wind directed more or less parallel with the run of the coast, although the wind mutes during the evening, the wind over the sea just off the coast will increase to compensate.

Small-craft sailors leaving port in the afternoon or evening to cruise coastwise must expect stronger wind during the night under these conditions.

Why does the wind speed change during the day?

It may help to know why these changes come about and it's mainly to do with convection currents.

With ascent three things happen:

1 Temperature falls
2 Wind speed increases
3 Air pressure falls

There are exceptions to these rules but this is what normally happens.

In the morning, as the ascending sun warms the land, up-and-down convection currents occur. This leads to parcels of faster wind coming down on sinking currents while slower wind near the ground gets taken up on ascending ones.

This 'mixing' of the air speeds up the wind you experience near the ground and, as the convection currents ascend during the morning, so the wind increases. As the ground cools during the afternoon and evening, it inhibits the convection and consequently the wind speed decreases.

As the sun sinks, so the convection currents fail and the mixing stops. Now the wind begins to die. During the day, visiting parcels of stronger wind have kept it going but now, deprived of these, it has only its own momentum to sustain it. As it collides with trees, buildings etc, it slows down and may even become calm.

Now a new situation appears. The air near the ground cools but the warm air taken up by the convection currents during the day is left above it.

What's an inversion?

Because it is normal for air temperature to fall as we ascend (Rule 1 above), a situation where the air is warmer above than below is the opposite of normality and we call it an inversion of temperature or just an inversion. Inversions may occur at all heights but it is the ones that appear during the evening and night that we will concentrate on.

Inversions have a profound effect on the day's weather. They make a kind of invisible blanket that traps layers of cloud (photo 7). In the evening they keep smoke or other pollutants close to the ground (photo 3). In the morning the sun starts convection currents and so holes are punched in the inversion. Until this happens, the wind may not wake up. The sign of the overnight inversion being broken is when you see heap clouds begin to appear.

The setting-in and breaking of inversion blankets makes the day's weather more staccato than Fig 2.1 would have you believe. However, on the whole, both wind speed and amount of cloud follow a daily rhythm when the weather is mainly settled.

Photo 7 These layers of lumpy cloud (stratocumulus) are typical in the evening. They may look dark but they are often cumulus clouds that have spread out under an inversion as the sun sinks. This is not, in any way, a threatening sky.

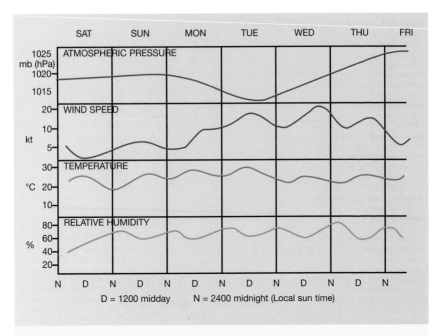

Fig 2.2 Here is an example of how the weather elements changed day to day over a six-day period when the weather was set fair. Note how the wind speed rose to maximum during the afternoon, slavishly following the temperature (except for Saturday), but see how the relative humidity did the exact opposite: being highest during the small hours and lowest during the afternoon.

We can best answer this question by taking an actual example of how the weather elements changed during a week of settled weather (Fig 2.2).

We see that for the first three days the pressure rose, then on Tuesday it dipped a little before rising again. Often, when pressure is high, these little falls in pressure do not make much difference to the weather except to make it more cloudy.

On Saturday, contrary to normal, the wind was lightest in the middle of the day but after that it did what we'd expect. It was highest after midday and lowest around dawn. The air temperature followed the expected diurnal trends but note that the relative humidity (RH) does exactly the opposite. It is just as if you had turned the temperature curve upside down. RH is highest in the early hours and least during the middle of the day.

Photo 8 Looking like the ribcage of a great air monster, this sky paints sea and sand with orange hues. Expect wind later.

What happens in unsettled weather ?

When strong winds come along and bands of rain chase each other across the country, the regular undulations in the weather elements shown in Fig 2.2 are not likely to be present. Yet, whenever it gets a chance, the weather tries to follow the diurnal pattern. The sun will try to see to it that midday is warmer than morning or evening. Over land even quite strong winds – up to gale force – tend to be strongest in the afternoons and least during the night. However, it can often happen that, say, a wind that is strongest around dawn will have become much lighter by midday as the pressure pattern changes.

Some things to avoid doing on summer evenings

Kite flying This needs some wind, and ascending currents help. In the evening the wind is beginning to die and the air is stabilising or even sinking. Often the air seems to be full of holes when a kite, apparently flying well, suddenly loses lift and drops out of the sky. Late mornings and early afternoons are best for flying kites.

Sailing (especially windsurfing and if you're a beginner) Evening winds are often fickle and may desert you altogether. If you've never sailed a boat or board before, don't choose evening to try it. This is especially so for wind-surfing where you need a fairly constant wind to lean against otherwise you feel you will fall over backwards. If you go out on an evening with light winds, make sure you are not going to be swept away by the tide if the wind deserts you.

3 clouds

What makes clouds and rain?

The answer is water droplets and ice crystals. All the air around us contains molecules of water vapour that only show themselves when the vapour molecules club together and form minute drops of water or miniscule crystals of ice. For that to happen they need help in the form of particles of dust, salt from sea spray, chemicals of various kinds and also diminutive ice crystals from clouds that are higher up. These little solid particles are called condensation nuclei and there has to be at least one at the centre of every ice crystal or water droplet that grows. Without them there could not be any substantial rainfall. In effect the air has to be a bit dirty before we can get clouds and rain (or snow).

The chances of water molecules in the air colliding and joining are incredibly small. They need something much bigger than themselves (condensation nuclei) to latch onto. Then they begin to form a water droplet and, as nothing succeeds like success, the bigger the droplet grows, the faster it grows. Eventually it gets big enough to fall out of the cloud as a raindrop.

At high levels in the clouds, despite the temperature being far below freezing, water molecules latch onto surrounding ice crystals and form snowflakes – in fact, much of the rain that falls started as snowflakes that have melted on the way down. So rain clouds have to be deep. They have to be the great, more-or-less solid wedges of layer clouds that are formed along warm fronts and occlusions (photo 16, page 42) or the towering shower or thunder clouds that populate cool, unstable airstreams or occur along cold fronts (photo 10, page 33).

The above is the way in which almost all rain occurs, but there is another way to get wet – drizzle. We get drizzle when droplets combine and so grow big enough to fall out of the clouds. For this the clouds do not have to be very deep. Much rain and drizzle over the foothills of mountain ranges comes from this coalescence of cloud droplets.

The cloud catalogue

Clouds can be divided into two kinds – those that are composed of water droplets and those that are composed of ice crystals. Sometimes both exist in the same cloud.

We are used to water freezing at 0°C (32°F) but in clouds water droplets exist at temperatures well below freezing point.

So ice-crystal clouds have to be the highest in the sky (photo 14). They are the ones that have names based on cirrus and their typical heights are around 30,000ft (9km).

Cirrus clouds have been associated with coming bad weather for millennia – maybe as long as man has been looking at the sky.

> Mackerel sky and mare's tails
> Make tall ships carry low sails.

This is a scientifically well-founded observation and we now know why it has always proved so beneficial to mankind. It is because the long teased-out banners of cirrus (photo 33, page 79) appear well ahead of the bad weather of depressions.

At first the names that have been given to cloud types may seem daunting but they can be easily understood if we note firstly their relative height and secondly their shape.

There are three height decks:

Low clouds form in the deck below about 7,000ft (2km) and their names do not have any height prefixes.

Medium clouds occupy the deck between 7,000 and 22,000ft (2–7km). Their names have the prefix alto and they are mostly water-droplet clouds.

High clouds exist above about 4 miles (6–7km) and they are usually wholly made of ice crystals. Their names are cirrus or are prefixed cirro.

There is nothing hard-and-fast about these height decks. In winter, ice clouds can come down into the low cloud bracket (photo 11, page 34), but on the whole these simple divisions will suffice. Everyone (including the professionals) has difficulty assessing cloud height with any degree of accuracy and you don't need to. You can usually recognise a cloud's height from its shape and name.

CLOUD SHAPES

Heap clouds are, in general, rounded lumps when they are small and their names are based on cumulus.

Layer clouds have names with a suffix based on stratus.

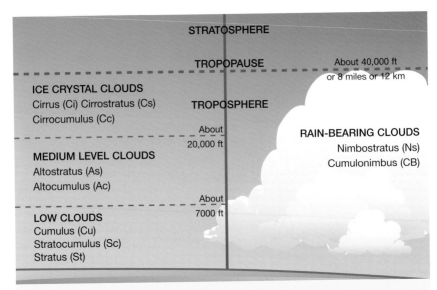

STRATOSPHERE

TROPOPAUSE

About 40,000 ft
or 8 miles or 12 km

ICE CRYSTAL CLOUDS
Cirrus (Ci) Cirrostratus (Cs)
Cirrocumulus (Cc)

TROPOSPHERE

About
20,000 ft

RAIN-BEARING CLOUDS
Nimbostratus (Ns)
Cumulonimbus (CB)

MEDIUM LEVEL CLOUDS
Altostratus (As)
Altocumulus (Ac)

About
7000 ft

LOW CLOUDS
Cumulus (Cu)
Stratocumulus (Sc)
Stratus (St)

Fig 3.1 The clouds tend to split into three decks but no cloud can go higher than the tropopause. The highest clouds – the ice-crystal clouds – exist between about 20,000ft (4 miles or 6km) and the tropopause (about 40,000ft, 8 miles or 13km). The medium-level clouds exist between about 7,000ft (1.3 miles or 2km) and 20,000ft. Low clouds occur below about 7,000ft. These height divisions are only averages and the tropopause itself is sometimes higher and sometimes lower than the 40,000ft quoted. The deep rain-bearing clouds span these decks.

HEIGHT AND SHAPE TOGETHER
In the **high cloud** bracket we have:

Cirrus (Ci) High, white, resembling locks of hair.

Cirrostratus (Cs) High, milky veil. Associated with ring haloes.

In the **medium cloud** bracket we have:

Altostratus (As) Sheets of layer cloud associated with coming rain.

Altocumulus (Ac) Rafts or islands of medium-size globules.

In the **low cloud** bracket we have:

Stratus (St) Low, often formless, fog-like cloud. Covers hilltops.

Cumulus (Cu) Low, rounded clouds, often associated with good weather.

Stratocumulus (Sc) Layers of heaps.

DEEP RAIN CLOUDS

In addition, there are the two cloud types that are possibly the most important in that they bring us most of our rain: nimbostratus and cumulonimbus. These deep rain clouds span the height brackets.

Nimbostratus (Ns) Deep layers of dark rain-bearing cloud associated with fronts.

Cumulonimbus (Cb) Towering heap clouds associated with showers and thunder.

More about clouds

High, water-droplet clouds (alto clouds) occupy the middle levels of what is called the troposphere – that part of the Earth's atmosphere where clouds can occur. Above the troposphere is the stratosphere and separating them there is an invisible 'lid' called the tropopause. Clouds cannot rise higher than the tropopause so the highest clouds (cirrus) are below the tropopause – when we see the anvil-shaped top of a thunder cloud (photo 10, page 33) we are looking at where the invisible tropopause exists – and all the normal weather processes go on under the tropopause.

The tropopause is merely a discontinuity in the way temperature falls as you ascend. It is where the temperature permanently changes either to stay much the same or even increase as you go higher.

LOW CLOUDS

Any low heap clouds that are more or less separated by gaps are going to be cumulus. Cumulus can be small, with their vertical extent not much greater than the height that their bases are off the ground. These are fair-weather cumulus (photo 6, page 11). When they grow deeper than this, they may be on their way to being shower clouds but they don't have to rain (photo 24, page 61). Once rain falls from them they are going to be cumulonimbus because the word nimbus means rain-bearing.

Much rarer than cumulus are the low, fog-like clouds called stratus (page 1). We get stratus under higher clouds when the weather is bad or it blows onto coasts if the wind is from a warm, wet quarter. However, in hilly or mountainous districts stratus often forms when the humid wind is forced up the hillsides and mountain-sides. It then shrouds the hilltops and, if you are caught in it, it will, as far as you are concerned, be fog. When cumulus clouds spread into a layer, they are called stratocumulus (photo 32, page 78). Meteorologically, things are not likely to change much when you have extensive stratocumulus.

MEDIUM-LEVEL CLOUDS

These are again either in heaps (altocumulus) or in layers (altostratus).

You can recognise altocumulus because it is often in rafts or islands around the sky and the globules are small compared to the familiar low heap cloud, cumulus.

Altostratus is a well-known cloud-type because it builds up layer upon layer before rain breaks out (photo 20). Everyone is familiar with the leaden look of the sky when rain is in the offing. Often you get mixed skies with areas of altostratus amid patches of altocumulus. Such skies occur when fronts have become old or when warm, humid air streams in from the subtropics.

There are two special kinds of medium-level clouds that often presage thunder. These are called altocumulus floccus (or floccus for short) and altocumulus castellanus (or just castellanus). With these clouds, the sky is often broken and chaotic and castellanus is so-called because it forms in lines with lumps that look like battlements sticking out of the top.

Altocumulus floccus (Ac.floc) These indicate approaching frontal systems with thundery outbreaks.

Altocumulus castellanus (Ac.cast) These little clouds with turret-tops may indicate thunderstorms later.

In hilly or mountainous districts, lens-shaped alto clouds called lenticularis are formed when the wind blows over the hill ridges. Lens-shaped clouds can also appear in the low and high brackets (photos 47 and 50).

Altocumulus lenticularis (Ac.lent) These lens or saucer shaped clouds do not indicate rain.

HIGH CLOUDS

These clouds are normally easy to recognise because they are composed of ice crystals and are pure white (this is compared to alto clouds, which are water droplets and show shadows). The kind of high cloud that runs ahead of deteriorating weather is cirrus and its trademarks are streaks, either hooked backwards or falling down the sky (photo 38, page 87).

Cirrus (Ci) High, white, resembling locks of hair.

Sometimes so many individual cirrus clouds are produced that they band together to form dense banners. This form often forewarns very bad weather on the way (photo 33, page 79).

Usually, when bad weather is in the offing and following the cirrus, a milky veil of cirrostratus covers the sky; this is easy to recognise as ring haloes form around the sun or the moon. Cirrostratus is the only cloud that forms these haloes (photo 15, page 41).

The third type of ice-crystal cloud is cirrocumulus. Don't worry about trying to recognise cirrocumulus. For one thing it is very rare and often you cannot tell it from areas of small altocumulus. Apart from possibly being the cloud type that is called 'mackerel sky', in weather lore it does not have much prognostic value, unlike its stablemates, cirrus and cirrostratus.

THE DEEP CLOUDS
The deep-layer clouds are nimbostratus (photo 12, page 36), while the towering heap clouds are cumulonimbus (photo 10, page 33). Between them, these two cloud types bring us almost all our rain. Nimbostratus is the cloud of weather-fronts and produces most rain in the winter, while cumulonimbus is the cloud of showers and thunderstorms and produces much of our summer rain. Nimbostratus is responsible for long periods of continuous rain, while cumulonimbus brings sharp showers, which sometimes may become local deluges. They bring snow showers in spring when you hope winter is over and may produce snow before autumn is done.

What makes clouds dark?
Clouds go from the whitest of white to jet black but what makes them like that?

Let's dispose of one thing first – white clouds are thin and therefore they won't rain. However, if there is a thin cloud layer below another, the one below becomes a shade of grey. The whitest clouds are those formed of ice crystals, ie the cirriform clouds that are highest in the sky. Layers of altocumulus clouds – often looking like ripples on the seashore – are frequently almost perfectly white when they are high and thin.

It is when water clouds thicken that they darken. There are two reasons for this. Firstly the sun cannot penetrate them so easily and secondly, in the myriads of cloud droplets there are 'dirty' condensation nuclei. As altostratus layers thicken when rain is on the way the sky progressively darkens and this is most marked when such clouds wing in from some polluted industrial area. When rain eventually breaks out, the deep nimbostratus cloud layers are the darkest of all.

The entity that can produce almost a blackout, and which causes the street lights to turn themselves on even in the middle of the day, is most often the slow-moving cold front (or occlusion) which has come to you via an industrial area. Now the cloud is polluted with smoke and grime from factory chimneys.

Summer skies

In late spring and summer, when the sun is high in the sky, its warmth can penetrate sheets of higher cloud and set off cumulus clouds underneath. So summer skies are sometimes at two levels, with cumulus under higher altostratus and altocumulus. However, the higher cloud shades the lower and gives the latter the appearance of being somewhat dark and threatening. Don't let it worry you. If the cloud layers aren't deep, then it is unlikely to rain.

We won't get this cloud-at-two-levels effect in autumn and winter because the sun is not strong enough.

Why aren't clouds deeper?

When fair-weather heap clouds (cumulus) cover the sky their bases are often flat and about 2,000ft (600m) up.

Each cumulus cloud is the visible sign of rising air currents (thermals) that are warmer than their surroundings. But the rising wet air cools at a fixed rate and eventually cools to what is called the dew-point temperature. This is when the rising air begins to form cloud. On days of fair weather, you will see that the bases of the clouds are all at the same level.

The air in the clouds goes on rising until it has cooled to the same temperature as its surroundings, when it stops rising. This is where the cloud tops are. The reason the clouds do not grow deeper is that they have met a warm inversion layer and cannot rise through it (photo 6, page 11).

Incidentally, whenever you look at clouds around the sky, at whatever level, they will not have grown higher because of an inversion (see page 13). Inversions lead to cloud layers at all heights and the biggest is the tropopause, which no clouds can penetrate.

It is when the airstream is unstable that humble cumulus clouds can grow to depths from which showers can occur. In this case, the rising thermal air only stops rising when it finds an inversion much higher up (photo 23, page 59).

To get an idea of the range of cloud heights, the grandest clouds of all – the thunder clouds of summer – grow to 6 or 7 miles (10 or 11km) high while the humble cumulus are usually only a twentieth of that.

Wrong forecast?

There has been hardly any cloud all day and it has been warm. The anti-cyclone or ridge of high pressure that is bringing the good weather is expected to last for a day or two yet. So why, if it is associated with bad weather, has cirrus cloud grown during the afternoon? Does this mean the forecast is wrong?

No! The cirrus that forms about the sky on hot days often means a continu-ation of the good weather. It has formed because the heated air has expanded upwards and pushed the highest levels into forming ice clouds. These clouds will be ragged and disorganised. They sometimes form loops and trailing filaments which lie in chaotic directions. It is when the cirrus forms in straight lines that you must expect a deterioration.

Photo 9 The ripples in this sheet of altocumulus cloud seem to mirror the sand ripples on the shore. This particular sheet of cloud is quite benign but there are signs of wind in the cirrus near the horizon.

4 about heat and cold

We need to keep our body temperature within certain bounds. An unclothed adult body loses some 90 joules of heat every second and if the outside (ambient) temperature is greater than the normal body temperature of 37°C the body will be absorbing more heat than it loses. This is when we feel hot.

To control the threatened increase in body temperature we perspire. The evaporation of sweat is a very efficient method of reducing the skin temperature and so the body naturally manages to keep its temperature within bounds. Only when the heat loss by evaporation is reduced by the kind of humidity you get in some thundery, tropical or semi-tropical conditions is there fear of over-heating.

When the ambient temperature is lower than normal body temperature we need to retain heat with clothing. To save feeling cold, the clothing must reduce heat loss to the 90 joules per second mentioned above.

If you access any forecast, the one thing they are sure to tell you is what the temperature is now or is expected to be in the future. But that is not the whole story, because as hairless human beings we have to allow for the wind chill as well. That means we also need to know what the wind speed is likely to be.

What is wind chill?

Whatever the temperature, if you expose your body to the wind it is bound to feel cooler. This is because the skin is always giving off moisture from its pores. This moisture has to evaporate and when a liquid evaporates it becomes cooler. Wind aids that process so wind helps to cool the body.

It has been calculated how cool or cold it will feel at a given air temperature and the wind speed is a certain value. You can assess the wind chill from Fig 4.1. Wind-chill temperatures are often given in forecasts but as we expect to feel colder when the wind blows, most people do not take much notice. However, it is when temperatures get very low – as they do in Canada, Alaska, Siberia and all the lands around the Arctic Circle – that a knowledge of wind chill becomes very important to reduce the risk of frostbite.

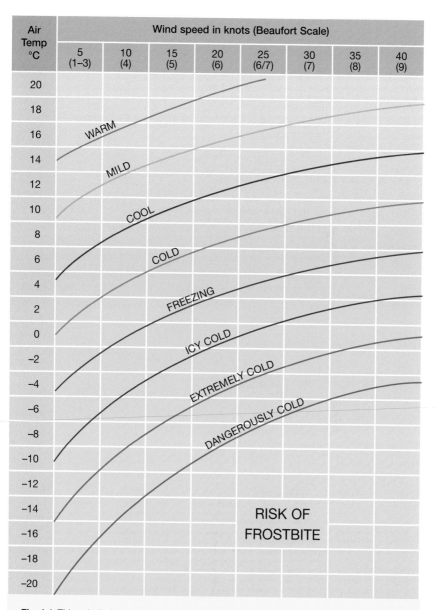

Fig 4.1 This wind-chill diagram shows how a combination of wind and temperature make it feel chillier than it would if there were no wind.

In summer, when we often need extra cooling, wind chill comes into its own. A wet body is cooled effectively by wind chill, which is one reason that we rush to get to the beach or into the pool.

The weather element that works against keeping cool is humidity. High humidity means that the net evaporation from the body is less or maybe nil. Even in temperate latitudes we sometimes get periods when extreme heat and high humidity come together. This is when people experience heat exhaustion or full-blooded heatstroke (see Why is it so hot? on page 32).

Do animals suffer from windchill ?

The short answer is that on the whole they do not and this is due to their furry coats. The air trapped in the fur insulates an animal from both heat and cold. Air is one of the best heat insulators so long as it says trapped – as it does in the confines of a fur coat. As long as a furry animal stays dry it will not suffer from wind chill but the insulating properties of the coat are much reduced when it becomes wet. Exceptions occur when animals are bred by humans. Despite growing winter coats, many horses used for competitions are clipped and need rugs. Small dogs, and in particular Mexican and other hairless dogs, need winter coats.

Fahrenheit or Celsius?

The Celsius scale of temperature is the one that has been adopted worldwide but Fahrenheit is still used, especially in America and it also clings on tenaciously in Britain. Each scale has its advantages and disadvantages.

FACTS ABOUT THE CELSIUS AND FAHRENHEIT SCALES

All temperatures below freezing are negative in the Celsius scale, whereas with the Fahrenheit scale they rarely go into negative in the temperate latitudes – it has to be the equivalent of -18°C before the F scale goes negative. This is a distinct advantage for people who do not like negative numbers. However, the Celsius scale is more scientifically based and most of the world uses it. Forecasters prefer the Celsius scale because the normal error in their temperature predictions is about 1°C whereas it would be 2° on the F scale, making the forecasts look more inaccurate.

Freezing point of water	0°C	32°F	
Boiling point of water	100°C	212°F	(1°C = 1.8°F)

Here are some useful comparisons

−5°C = 23°F	Icy	5 = 2 + 3
0°C = 32°F	Freezing	
+5°C = 41°F	Cold	(5 = 4 + 1)
+10°C = 50°F	Chilly	(both end zero)
+16°C = 61°F	Mild	(1–6 and 6–1)
+20°C = 68°F	Warm	(maybe easier to remember 21° = 70°)
+25°C = 77°F	Hot	(2 + 5 = 7 and double it)
+30°C = 86°F	Very hot	
+35°C = 95°F	Too hot	close to blood heat

You will not be far out if you adjust the above up or down by 1°C = 2°F.

Why doesn't my thermometer give the right temperature?

If you want to be fairly sure of what the true temperature is then you have to site your thermometer properly. My garden faces north and I have three outside thermometers. They are called, for reasons that will become apparent, Stevenson, Rocket and Bathroom.

You can only expect to get temperatures that you can compare with official ones if you have your thermometer in one of those white, louvred boxes called a Stevenson Screen. I have one in the garden and Stevenson gives me an accurate temperature, often less than I think it is in summer and more than I think it is in winter.

I also have a thermometer bought from a garden centre on the back, north-facing wall where unfortunately it can see the concrete patio and the back lawn. This thermometer consistently reads high whenever the sun heats the patio etc and so is called Rocket. Unlike Stevenson, it picks up heat radiated from the patio and lawn and so over-reads. Rocket is at 4ft (1.2m) from the ground at about the standard height, but up on the back wall of the house, outside the bathroom window, I have Bathroom, a distant-reading electronic temperature element cached under a white-painted plastic cover with holes in it. This gives me the temperature at an 'altitude' of about 20ft (6m).

With Stevenson it does not matter what time of year it is, I can still believe its temperature reading. Not so Rocket and Bathroom.

From the spring equinox (when the sun rises due east and sets due west) right through the summer until the autumnal equinox the sun will, in the early morning and the evening, be north of the east–west line and so Bathroom will be struck by the sun's direct rays and over-read. It will be like that until 6 am local sun time and also after 6 pm local sun time. So I can believe my bathroom thermometer much more in the winter than I can in the summer. Rocket is shaded by trees in the morning and by the house next door during the evening so the above does not apply.

My latitude is about 52°N and the sun rises some 40° north of east and sets 40° north of west at the summer solstice, 21 June. If you are further north there are longer hours of daylight; further south there are less. Yet on whatever latitude you live, the sun will always rise and set due east and west at 6am and 6pm local sun time at the equinoxes (21 March and 21 September).

That's what you have to take into account in the horizontal. What about the vertical? Stevenson is the regulation 6ft 6in (2m) above the ground. Rocket is a negligible bit lower than that, and Bathroom is much higher.

Let's explore the extreme conditions. On cold, calm nights, the air near the ground becomes layered so that the coldest air temperature is near the ground and it gets warmer as you go up. So Bathroom will read higher than the other thermometers and sometimes, especially in winter, quite a few degrees higher than on the ground. So it is no good relying on Bathroom to tell you when it is freezing on the ground. On the other hand, on hot summer days the highest temperatures will be near the ground and it will get cooler as you ascend. This time Bathroom will read lower than Stevenson. When a strong wind blows, all three may have much the same reading.

In other situations – for example, in the confines of a yacht – it will be very difficult to find somewhere for a thermometer that gives the right reading. If you have an electronic temperature read-out, where is the temperature-sensing element? Can it feel efflux from the galley or see radiation through the coachroof? You may be lucky enough to have a sea-temperature sensing element as well, in which case you can compare the two and see what the prospects of sea fog are (page 108).

Why is it so hot?

On the best of summer days the temperature may rocket during the morning and you expect it to reach record levels. But it doesn't. Sometime during the late morning or early afternoon the quiet conditions of the early morning will be replaced by a breeze. This levels off the temperature – may even make it fall a little – because convection currents have mixed the heated air near the surface with cooler, faster-moving air from higher up. In hot weather you'd expect the convection to mix up the air to a greater depth and keep your air cooler. So how do we get temperatures into the 90s F or the middle 30s C? The answer lies in the anticyclone that is producing the fine summer weather.

ANTICYCLONES IN SUMMER

Anticyclones are regions where air is subsiding. Just as rising air cools, so sinking air heats up. This way strong anticyclones produce layers of very warm air not very far above the surface, say 2,000ft (600m) or so up. This inhibits the thermal currents that would mix the surface air layers. The result is that a thermal 'lid' (a strong *subsidence inversion* of temperature) keeps the sun's heat trapped in a layer that is not very deep and the result is that temperatures climb towards record levels.

In coastal lands or relatively small islands like Britain you can only get extremely high temperatures with the aid of these subsidence inversions and it has to be a strong one relatively near the ground. If you want to see where the inversion is then look at the western horizon just after the sun has set. You will often see a grey-blue layer near the ground which is visible because of pollution trapped below the inversion.

In continental situations well away from the sea, it is much more likely that extreme temperatures will be reached but it still needs the intervention of a strong anticyclone. In the unprecedented European heatwave of August 2003, France was worst affected and some 14,000 people died of heat-related illness, well above what would have been expected. Paris had its hottest night ever on 11/12 August, the thermometer not falling below 25.5°C (78°F). In the countries surrounding France a further 10,000 deaths occurred and the extreme heat did not relent for a full ten days. The culprit was an immobile anticyclone sprawled across western Europe.

Photo 10 The great anvil-head of a thunderstorm arches over the coronet of cumulonimbus cloud that always surrounds storms. The top of this cloud is some 6 miles (10km) high and the temperature up there is perhaps −40°C.

ANTICYCLONES IN WINTER

When a strong but sluggish anticyclone sets in over land in the middle of winter, its hold is very difficult to break. In summer the high sun gets to work on the cloud tops and evaporates them away, so summer anticyclones are largely sunny. Not so in winter. The sun does not have enough power to make an impact on the tops of the oceans of stratocumulus cloud that fill the depths of winter anticyclones. The strato-cumulus layer becomes trapped below a low subsidence inversion and anticyclonic gloom sets in. The clouds do not break for days – in extreme cases weeks – on end and all the pollution below them is trapped and cannot escape. The result may be smog ie smoke-fog which is deadly for those with respiratory conditions.

Even young, fit people find the cold, dismal days very damaging to morale and many people suffer from the condition known as SAD (seasonal affective disorder).

How do winter mornings get so cold?

For the earth to get cold overnight it has to lose heat by radiation. That means there must be clear (or nearly clear) skies – again it may well be that you are under an anticyclone. Very often the air is cold already and if the ground is colder the air loses heat to it by conduction. The result is ground frost.

Photo 11 In a very cold airstream the fallstreaks, usually associated with cirrus cloud, can come down almost to the ground. This unusual event coincided with the passing of a cold front.

Right-now weather?

It is a warm summer morning – hardly any wind and not a cloud in the sky. The thermometer is rocketing. Can we expect some relief?

The normal increase in wind with the day (page 10) should give lower temperatures but if you live, say, within 10 miles (16km) of the coast you ought to get a seabreeze soon. When you are under a subsidence inversion due to an anticyclone, then seabreezes are inhibited – they only travel slowly from the coast and may not arrive until the afternoon. However, if you see some puffs of cumulus cloud around the sky, seabreezes are aided by convection and travel inland faster and further. They also blow more strongly so places as far as 30 miles (50km) inland may feel their effects.

The other factor that contributes to a ground frost is little or no wind – another attribute of anticyclones. However, if there is wind in the evening or night no air frost may occur because the wind mixes warmer air from up above with the colder surface air and so keeps the temperature of the latter above freezing. There may still be a ground frost, however.

If the airstream is near or below freezing already then clear skies usually mean a very penetrating frost – the kind that freezes water pipes and unprotected car radiators. The wind need only be a few degrees below freezing – but the wind chill will be severe.

Sometimes, when the air is humid, the evening fall of temperature may induce radiation fog. However, fog is an enemy of frost as it provides a kind of blanket to insulate the ground. Conversely, frost is an enemy of fog because if the moisture is frozen on the ground it cannot contribute to fog. When I was a forecaster we had a rule of thumb: If we fog we don't frost and if we frost we don't fog.

However, on some occasions you can get freezing fog that makes for a fairyland of hoar-laden trees.

5 rain

Interestingly, almost all of our rain starts as snow in very high clouds. Continuous (and often non-continuous) rain falls from the deep cloud wedges called nimbo-stratus, so these clouds are associated with weather-fronts.

It may be very hot and humid before a thunderstorm but up in the higher reaches of the thunder clouds it is very cold – so maybe just a few miles (or km) above your head there is snow. It's a fascinating fact that we only get big raindrops because they start as snowflakes. Snow has a parachute effect and allows the incipient raindrops to fall slowly to lower regions of the clouds before they melt. Thus they only fall from a relatively low altitude. Look at the side of some great shower cloud and you will be able to see the characteristic bulges and streaks where the snow is falling.

Photo 12 The weather is clearing and you can see the depth of these nimbostratus clouds, plus all the scud below, so you get an idea of the wind strength from the sea.

Why is it raining?

The usual answer is that the wet air in the clouds is ascending. It may ascend slowly – as along fronts – or rapidly, as in a big shower or thunder cloud. The slow ascent leads to continuous rain while the rapid ascent leads to showers.

In Chapter 7 we will look at the way that clouds form along warm and cold fronts as well as the amalgamation of the two – an occluded front.

We associate layer clouds with continuous rain (or drizzle) but sometimes the rain from sheets of layer clouds comes in spurts with dry periods between. This kind of showery rain is called *intermittent*. Another form of intermittent rain is described as 'thundery showers' and to find the reason for this we have to take to the skies and look at the top layer of the clouds. As we saw in the previous chapter, when thundery showers occur, the top of the cloud layer will have big heap clouds sticking out of it. These showers can be locally heavy and they often shoot bolts of lightning between them with accompanying thunder which, because it is so high up, rolls around the sky making a lot more noise than the conventional thunderstorm. Occasionally there is enough electrical energy to create a lightning flash to the ground.

Intermittent rain is normally due to the growth of shower clouds within the main cloud layer and they may also protrude into the blue sky above.

Do you live in a rain shadow?

If you live just to the east of substantial hills or mountains, you will often be drier than others who live further away. You are living in an area known as a *rain shadow*.

Clouds over the sea may be able to take up moisture from below to make up for what they lose by raining. However, when they drift ashore their moisture content is constant. However as the clouds climb the foothills and the mountain slopes they are forced to drop much of this water. So people who have mountains to the west enjoy a much drier climate than those who live on the western side, especially if facing the sea.

You don't have to have mountains in the direction of the prevailing wind to experience a much more equable climate than others not far away – just moderate hills will be enough. The coast of Sussex in southern England is well known for its balmy climate. The helpful hills in this case are the South Downs, which protect the coasts from the colder, showery north-westerly or northerly winds.

Why isn't drizzle just thin rain?

Don't confuse rain and drizzle because they are produced in entirely different ways. Drizzle could never have started as snow as rain does. Drizzle comes from low stratus clouds – which resembles fog near the ground – and consequently visibility is often poor in drizzle.

Photo 13 The cloud is low enough to cap the high ground. Stratus is the enemy of hill-walkers.

Drizzle is probably at its most intense when wet air is ascending the slopes of hill and mountain ranges – local conditions are then at their worst. The visibility can be just like dense fog but you get saturated by the persistent onslaught of this mass of tiny drops. Meteorologically speaking, the clouds that produce rain are not those that produce drizzle. Drizzle goes with very moist, relatively warm, air. Low stratus cloud will appear in bad weather systems, especially at sea, and it will cover headlands as well as sometimes becoming fog when the air is lifting over rocky coasts.

Radar and rain

Suitable radar can get echoes back off raindrops, hail and snow. Rain radars may not, however, be able to detect drizzle because of the small size of the drops. In Britain radar covers the whole of the country and reveals things we could never have divined from ground observations.

The best way to follow the progress of rain belts is to access radar plots on the Internet. You can then see where the rain is actually occurring and how intense it is. The British Met Office gives a radar plot every half an hour and allows you to run a sequence that will show how the rain is moving, as well as whether it is getting more or less intense. The plots are about half an hour out of date when you receive them but when we are dealing with continuous rain that is not a great drawback.

Weather fronts coming

Most of the rain and snow we get comes from *fronts*. These are bands of deteriorated weather where warm air overrides cold (warm front) or cold air drives in under warm (cold front) (Fig 7.1).

Warm fronts sometimes bring long periods of more or less continuous rain while cold fronts often start with heavy rain, which becomes less with time. The processes of weather within depressions lead to an amalgamation of warm and cold fronts called *occluded fronts*. Often occlusions produce less virile weather than warm or cold fronts but sometimes they are responsible for long periods of overcast skies and rain and drizzle when they get stuck over an area.

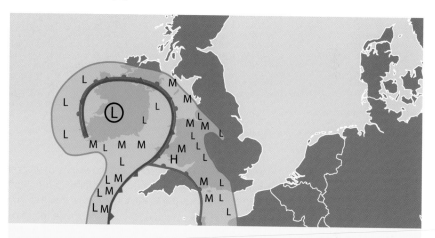

Fig 5.1 The rain-radar plot during mid-afternoon on 26 March enables us to sample the intensity of rain around a depression and its fronts. We find big variations in the intensity. Here, the green area shows the extent of the rain-bearing clouds. There is a depression centred over western Ireland and warm and cold fronts down over the Western Approaches. An occlusion bends back around the low centre. L = light rain; M = moderate rain and H = heavy rain. We see that moderate areas are interspersed by light ones and (in this case) only one area produces heavy rain, but this is unusual. Heavy rain often occurs along cold fronts, for example. At other times 'holes' with no rain will appear.

If you can access radar rain plots (Fig 5.1) on the Internet, they will show how variable rainfall can be – so weather forecasts, which cover large areas, cannot say exactly when it will rain on your little patch or how much rain you will get. We know by experience that when the sky gets a certain lowering look, rain is likely very soon – but what about a longer-term forecast?

Is rain coming?

Often, long before there are any other signs of coming rain, there are wisps of cirrus cloud invading the high sky (below).

Photo 14 Cirrus cloud coming in before a depression. This is typical of this high, ice-crystal cloud with its wisps and lines. It is a harbinger of deteriorating weather.

They often creep in unnoticed because we are enjoying the fair weather in the ridge that precedes the coming deterioration and don't trouble to look above the fair-weather cumulus clouds that populate the sky.

Once you are aware of the high ice clouds, you suspect that rain will follow – but it is not always certain. Cirrus clouds appear for many reasons and they even appear on hot summer afternoons when there is no chance of it raining. If you want to check that it is going to rain, you have to keep an eye on the sky. What you are looking for is the cirrus getting denser and more extensive.

A sure sign of trouble coming is when the wind backs into the south, in which case the cirrus will be coming in from the west.

> A veering wind will clear the sky
> A backing wind says storms are nigh

(Veering means shifting clockwise or moving with the sun. Backing means shifting anticlockwise or back against the sun.)

One important thing to remember is that the wind direction aloft is rarely going to be the same as the one you feel at the surface. In fact when weather is really on the change the winds that are carrying the cirrus clouds are likely to be at right angles to the surface wind.

Some local signs

I learned some good weather predictions at an early age from my mother. Our house was built in the 1920s and it had sash windows. We lived in a village close to the Sussex-Hampshire border and, at the back, we looked out southwards over the marshes of Chichester Harbour, so there was little protection from southerly gales that gathered across the open vista towards the English Channel and Isle of Wight. It was when the back windows began to rattle that my mother predicted rain was not far away. The chaotic gustiness of the wind indicated that bad-weather clouds were gathering. So all those who have double-glazing have lost a reliable weather predictor.

Another of her items of weather wisdom was to predict rain when the smoke from the chimney blew downwards rather than upwards. Also sailors are always taught that funnel smoke blowing down is a sign of coming trouble. My mother's observation was correct because there is much sinking air under gathering warm fronts, which is why you rarely find lower clouds below the high clouds.

My mother was also fully aware of what a ring around the sun or the moon meant and she instilled a feeling for weather that has lasted me all my life.

Why is the sun in his house?

Following the cirrus will come the ice-crystal cloud that spreads a milky veil across the sky causing the ring halo effect round the sun or the moon. This layer of ice cloud is called *cirrostratus*. The weather lore says:

> When the sun is in his house it will rain soon.

Statistically there is a better than 75 per cent chance that when you see a halo, rain will follow so the lore is well-founded.

Photo 15 There are many apparent shapes made by the sun in cirrus clouds. They include circles, arcs and crosses but by far the most well-known is this kind of solar halo. It is also the most useful, for it warns of rain to come on most occasions. If cirrus changes to this cloudy veil (cirrostratus), rain is almost certain to follow in a few hours' time.

Photo 16 Nimbostratus of a warm front clearing, with rain curtains hanging below its base. Looking through, we see the clouds that will follow the front. They look fairly benign, so we can expect it to clear up.

These cloud types are being spawned above a warm front (or sometimes an occluded front) but not all fronts are the same. Very active ones will show all the different cloud types that appear with fronts, but less active – maybe older – ones can pass without a halo appearing. In every case, however, cirrus will be present even if later, as the clouds gather, you cannot see it.

How long before it rains?

The halo phase passes and more substantial clouds appear. These soon thicken to shroud the sun and make it gradually disappear. However, it will probably be some time – usually hours – before these water-droplet clouds grow deep enough to rain. The darker, lowering cloud layers are called altostratus and they are often featureless.

Sometimes, especially with occluded fronts, there can be brief breaks in the cloud sheet or the rain can appear in a matter of an hour or so. There is a useful bit of weather lore here that helps gauge how long it will be until it rains and also how long the rain may last.

> Long foretold – long hold
> Short forecast – soon past

Should I reach for an oilskin or an umbrella?

Nature has ways of telling us what she is up to. A case in point is when it is about to rain. The thickening layers of altostratus that are making the day more gloomy do not normally have any clouds beneath them. However, just before it begins to rain, puffy lumps of cloud begin to cross the cloud base. This new cloud is called *pannus*.

It is forming because thin rain is falling from the clouds above and as this evaporates it moistens the air below the cloud base sufficiently for cloud to form there. If you see the pannus forming, you can be certain that the rain will reach the ground within a few minutes. Very soon the real cloud base, from which the rain is falling, will have been lost in a total cover of this amorphous pannus and the apparent cloud base may only be a few hundred feet off the surface.

When will it stop?

Usually it will go on raining now until the warm front passes. This normally takes hours but you can use the bit of weather lore above to get a better idea of how long it will be. Rain belts will not rain continuously at the same rate for hour after hour. They will rain lightly at first and then become moderate or even heavy, but the moderate or heavy rain will not occur all the time, nor everywhere. There may even be some periods when it hardly rains at all (Fig 5.1).

What will happen when the warm front passes?

Firstly the rain will stop. There is often a bright or even clear patch behind the front, but do not trust it – very soon the clouds can be down near the ground again. The air will be warmer and more humid and there may be some more rain, although drizzle is more likely. The risk now is fog, especially on coasts or over the slopes of hills. If it is the summer half of the year, it can be hot if the clouds break.

Photo 17 This sky is typical of those that look threatening but are not. A hard, knobbly base means there will be little or no rain, although do not be too complacent – sometimes patches of a sky like this will go fuzzy and produce a shower.

Continuous rain? Very rarely

It is very unusual for a rain belt to continue without a break for hours on end. It does happen, when fronts are new and vigorous and come in straight off the ocean or over hill slopes facing the wind. Otherwise, as the rain belt travels inland, some parts become drier than others so there are dry periods interspersing the wetter ones.

Experience shows that rain from warm fronts and occlusions starts light and becomes heavier. It differs from cold-front or showery rain, which starts heavy and then gradually tails away.

After a period of being moderate, the rain usually goes back to being slight before a drier interlude comes along. (This is well illustrated in Fig 5.1, page 39, before the rain clears.) However, you cannot be sure if you are near a clearance and the drier interlude is what it seems – it may not last long. You may not even notice the periods when the rain has all but ceased as the air is wet, the ground is wet and high humidity is the order of the day. Sometimes, the interlude when it is not raining is filled with drizzle.

However, there is another reason that there are real gaps in the rainy periods and this is multiple fronts. Fronts, which are represented by a single line on weather maps, are sometimes two, or even three fronts chasing one another. They are usually quite close in time but there is a definite dry period between them. It usually remains cloudy but occasionally glimpses of blue sky appear before the weather closes in again. You cannot be sure that a front has cleared until you can see a great wedge of blue sky following in from windward.

Is it about to rain?

The forecast was for rain and you are waiting for it to arrive. It seems to be imminent as the sky becomes lowering and considerably darker.

Suddenly things change – the sky lightens – there may even be a glimpse or two of the sun. This makes you think it is not going to rain after all.

The advice is not to trust it – there is often a temporary break before the main rain sets in. This is because the air in the coming clouds must be ascending to make rain. But if air is ascending in one place it must be sinking in another. This sinking occurs ahead of the rain band proper and often makes for a temporary break.

Summary of signs of rain

IN THE LONG TERM
(MAYBE SIX OR MORE HOURS)

Increasing high clouds (cirrus, altocumulus etc)

Fair-weather clouds dispersing or spreading out

Exceptionally good visibility

Haloes about sun or moon

Grey layers of cloud coming in

Wind shifting direction anticlockwise (backing)

Barometer falling

IN THE SHORTER TERM
(LESS THAN SIX HOURS)

Sun disappearing

Sky becomes overcast with few, if any, breaks

Sun dogs appearing in increasing high cloud

Wind increasing and maybe becoming more gusty

Smoke blowing down

Insect-seeking birds (swallows etc) fly low, particularly over water

Humidity increases

If there is no dew in the morning, rain is likely during the day

Barometer falling faster than before

IN THE VERY SHORT TERM
(MAYBE LESS THAN AN HOUR)

Sky overcast and lowering while darker lumps of fuzzy-edged cloud appear below it

A visible wall of deep clouds to windward often mean squalls and rain

Signs to windward of rain stalks

Certain flowers close up eg daisies and dandelions

Flying insects disappear

6 changeable weather

The weather of the temperate latitudes is nearly always changing. We do get some long periods with day after day of settled weather but these are the exception rather than the rule. However, we do not have to go far as the weather flies – for example, to the Mediterranean in summer – to get settled weather for weeks or even months.

Atlantic Europe is more likely to get settled weather in the summer because of the way a ridge of high pressure from the semi-permanent Azores anticyclone builds out north-eastwards at this time of year. Even so, depressions are never far away and – given a chance – will invade what we might otherwise hope will be a fine summer.

Go deeper into Europe and settled weather is more usual than it is over the lands bordering the Atlantic seaboard.

Statistics show that the lands surrounding the English Channel and on across Europe, some hundreds of miles (or km) either side of the 50°N parallel of latitude, are the most changeable, both summer and winter.

To the south of this zone, across Spain, Italy and the Balkans, things are less changeable with a bias towards spells of better weather, while to the north it is more likely that you will get more changeable spells.

There is a region of the Atlantic south of Iceland where the chances of getting depressions are highest. This does move a bit northwards during the summer but it is still there. 'Pressure is low to the south of Iceland' is a familiar phrase in the shipping forecasts at most times of year. Scandinavia is another area where the weather is very changeable in winter but is more likely to be cyclonic in summer (see below). Although most people may not believe it, the northern Mediterranean is also a region that is mainly cyclonic in winter. This zone stretches from the Gulf of Lions across southern Italy and into Greece and Turkey. In summer it changes entirely to being mainly settled.

North America is such a vast and varied region that its vagaries cannot be adequately covered here. The mountainous backbone that threads down the western side separates an oceanic climate like that of Atlantic Europe from the central areas with a more continental climate that is much drier. The eastern seaboard has some wide divergences in weather types with quite a lot of snow in winter and high humidity in summer. The states surrounding, and inland from, the Gulf of Mexico suffer from big thunderstorms and consequent tornadoes, as well as being in the path of late summer and early autumn hurricanes.

Canada breeds its own inland depressions in the region of Alberta, which feed changeable weather to eastern Canada and New England.

What is cyclonic weather?

Cyclonic weather is generally poor weather. It is what we expect when depressions are near. However, the word *cyclonic* has two distinct meanings. Meteorologists talk about the isobars being 'curved cyclonically'. Isobars can be straight or else curve one way or the other. In Fig 6.1 we see that over the northern North Sea, as well as over the states to the north-east of Italy, the isobars are curved cyclonically. This means that the isobars are curved in a way that attempts to enclose low pressure. They do not have to enclose it fully – they must just curve that way.

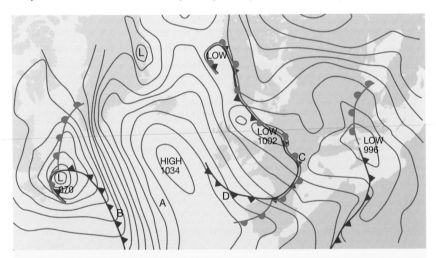

Fig 6.1 The weather map for 0600 GMT Saturday 5 March 2005. The green lines are iso-bars – joining places of equal pressure. Red lines with bumps are warm fronts; blue lines with spikes are cold fronts, while the two superimposed on one another are occlusions. The dominant feature is the anticyclone (high) covering most of the North Atlantic and a low between Scotland and Scandinavia. At A isobars are weakly cyclonic; at B they are strongly cyclonic; at C merely cyclonic while at D they are, again, weakly cyclonic.

It is a fact that when the isobars are curved cyclonically, the weather under them is not so good. This could be why your weather has recently deteriorated. Maybe a little later, when the pattern of cyclonic isobars moves on, it will get better again.

We also hear the term cyclonic in shipping forecasts. In this case, when the wind directions are described as cyclonic, it means that they are going to shift in the way that we expect when a small low tracks through the sea areas in question. The changes in wind direction and speed cannot be adequately described as they will depend on where you are compared to the centre of the low, so the forecasters rely on your own knowledge of what is likely to happen in this cyclonic situation.

What is anticyclonic weather?

Anticyclonic weather is generally good weather. When highs sprawl their isobars over you, winds tend to be light, clouds disperse and sunshine is the norm. At least, it often is in summer. At other times anticyclones can be very cloudy and, as Fig 6.1 shows, over the mid-Atlantic, winds can be very strong around their periphery.

Again, just as with cyclonic weather, there are two meanings to the word *anticyclonic*. There is the one described above and then there is the case of anticyclonic isobars.

When the isobars curve so as to try to enclose higher pressure, we say they are anticyclonically curved. In Fig 6.1 the isobars are strongly anticyclonic over Spain and they indicate that the air there is sinking and the weather should be mainly fine. However, we must always allow for the fact that anticyclones are often cloudy. The good weather is then above the clouds and being enjoyed by airline passengers.

Because anticyclones are big sprawling things, anticyclonic is a term you will never hear in a shipping forecast when referring to the way winds are going to change.

Photo 18 As the clouds of a coming front lower and darken, you expect it to rain – but when? Learn to recognize these lumps (pannus) under the cloud base, which form just before rain breaks out. It rained ten minutes after I took this picture.

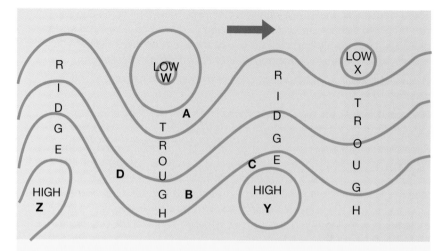

Fig 6.2 An idealised situation when, during unsettled weather, troughs of low pressure are interspersed by ridges of high pressure. The sequence is moving eastwards as a whole. With fairly large variations, this is what creates the changeable weather patterns when rainy (or snowy) periods are interspersed with fairer periods. The sequence is broken when a big anticyclone, like the one over the Atlantic in Fig 6.1, builds and blocks off the depressions.

What are troughs and ridges?

In a run of changeable weather, lows and highs chase one another from west to east around the temperate latitudes (Fig 6.2), so the isobars are definitely cyclonically curved at A and they are also cyclonically curved at B, but not so strongly.

Again, they are anticyclonically curved at C but also at D. The way the isobars curve has a great effect on the weather.

Meteorologists have shown that if you are under cyclonic isobars then your weather will be worse than expected, while under anticyclonically curved isobars it will be better. This is because air, where the isobars are cyclonic, must on the whole ascend, while where it is anticyclonic it must sink. Rising air leads to cloud and rain or snow, while sinking air leads to rain dying out and clouds dispersing.

Because of this, near a low centre (as at A) we expect overcast skies, rain and drizzle, but even at B there is more chance of cloudiness; showers are more likely and will be heavier. On the other hand, due to the sinking air, we can expect clear skies at C near the high centre but the air must still be sinking over D because of the way the isobars curve. So while it may well be more cloudy at D, the weather is most likely to be fair and the chances of rain are lessened.

These rules apply even to the local weather map. There may be no depression near you but if the isobars are cyclonically curved (ie curved as if trying to enclose low pressure), chances are that it will be cloudy, there may be rain (or snow) you did not expect and it might be windier or gustier but not necessarily colder.

Groups of isobars (as at B) form a *trough* and troughs are renowned for deteriorating weather.

With anticyclones we expect better weather – we may not always get what we expected but on the whole anticyclones bring fair, or even fine, weather.

Stretching out from high centres we get *ridges of high pressure* and much of our fair weather comes from ridges, especially in the summer. Because of the way the isobars curve, a ridge is a region where air must sink from very high up. If rising air leads to cloud and precipitation, so conversely sinking air leads to clouds being eroded with much less chance of any rain or snow. However, as we've already seen more than once, ridges and anticyclones generally can often be very cloudy and go on being cloudy for days on end.

Photo 19 When a cold front passes, the clouds lift in tiers until the last cloud to pass is cirrus, often followed (as here) by blue sky. The front is moving away to the left and the clear slot will open up the ground to the sun until it is warm enough to initiate thermals and so build heap clouds (this does not apply to the sea). Often, the clouds in the rear of cold fronts get big enough to produce showers.

What's going to follow a low?

However foul the weather may be now, we know it cannot go on for ever. Hidden in the bad weather is a forecast that, sometime in the not-too-distant future, things will improve. It might take a day or two but if the weather is particularly changeable with strong winds and blustery showers, then better weather is bound to turn up. We see in Fig 6.2 how this must be because if we are in the low X, and all this run of weather is moving as a whole eastwards, then a ridge will follow. The better weather may not last long because there is another depression (low W) waiting in the wings to push the short spell of good weather away.

You can believe the forecasts in this kind of trough-ridge-trough-ridge situation because one thing the meteorological computers do well these days is forecast where depressions are likely to be and how they will move.

What's going to follow a high?

While we may not like the idea, good weather regions move away. They may do it quickly – in a day or so – or they may take weeks, but eventually low weather will be back. Depressions throw out harbingers of their coming in the shape of characteristic high clouds and changes in the wind. Fine weather heap clouds that disperse or spread out into layers may also tell you that deteriorating weather is in the offing. These changes will be covered in the next chapter.

Does a depression make you depressed?

When the weather is cyclonic, ie due to depressions, then from a human point of view we are also depressed. To some people it may not matter but weather sensitive people are on the whole more at risk from having accidents when the weather is cyclonic. This means that it might be as well to look at the weather map for the day and see if the isobars curve over your area in the way described on page 49. The way the isobars curve has a direct effect on the weather and so can have a direct effect on you. Cyclonic means that more air is rising than is sinking in the region where the isobars curve in the sense of trying to enclose lower pressure.

Rising air leads to cloudiness and precipitation (rain, drizzle, snow, sleet) and so the general conditions conspire to increase the chances of making a nasty error of judgement when the weather is cyclonic.

7 frontal weather

In previous chapters we have looked at the kind of clouds and weather we experience when warm fronts are approaching. Now we want to look at the weather cold fronts and occlusions. The diagram Fig 7.2 is a cross-section of the clouds to be expected when a depression passes. It shows how the rain clouds build up when a warm front is approaching and the reverse sequence when a cold front passes. But the diagram Fig 7.1 shows that normally a warm front that's moving away must mean, some time later, a cold front to come.

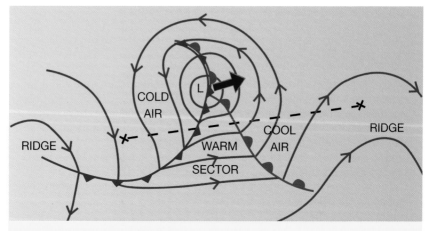

Fig 7.1 A text-book depression will look like this, with a ridge of high pressure in front of it and behind it. This depression is some days old but is typical of those that invade Atlantic Europe. It was born over the Atlantic by a meeting of warm moist air and cold moist air. Now the warm air is confined to a triangular region called the 'warm sector' and the rest of the warm air is above our heads, forming the fronts. The depression has begun to occlude; where that is happening, no warm air exists on the ground.

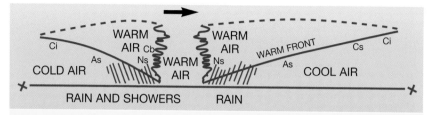

Fig 7.2 A vertical cross-section along the line X–X in Fig 7.1 indicates the deep wedges of cloud that build ahead of the warm front and behind the cold front. In the thin edge of the warm front wedge we have cirrus cloud and the other cloud types follow as the whole system moves eastwards.

How do I tell if a cold front is imminent?

This is a difficult question to answer because, unlike warm fronts, cold fronts often arrive without much warning (photo below). However, the air will be relatively warm and humid so it will almost certainly be cloudy, and if this cloudiness opens up and reveals some clear blue sky, suspect that a cold front may be in the offing.

Photo 20 Old fronts will have many of the cloud forms of new fronts but they are often broken up into islands and bands of cloud. The clouds will normally build up relatively quickly compared to a young warm front. The cirrus (top right) changes to thin alto-stratus and then to a dark band of thick altostratus but it is not wholly solid as the sun can peep through a gap.

Photo 21 Not all great masses of cloud stretched across the wind are fronts – some are troughs, which look rather like cold fronts and it is often difficult to tell the difference. Perhaps it does not matter – the trough will bring a period of showers as a cold front will, but not usually a veer of wind.

So among the signs are:

- A break in cloudiness that frequently comes just ahead of a cold front.
- A wall of heavy clouds bearing down on you. This wall will have the anvils of big shower clouds or thunderstorms sticking out of its top (photo above). If you do get the break in cloudiness then this is how the leading edge of a cold front will look.
- An in-depth forecast from a TV station or on the Internet showing a chart with the symbol for a cold front on it, or a mention of its imminent arrival.
- A shipping forecast or an inshore-waters forecast warning that a cold front is expected although rarely indicating when.
- The latest rain-radar plots off the Internet indicating the onset of the band of heavy rain and showers that characterise a cold front. Also look at the plot of lightning strikes (Sferics) that can be found on the Internet. This will show if there are any thunderstorms mixed in with the other clouds of the front.

What will happen when a cold front passes?

- Firstly it goes squally and there is heavy rain and showers.
- Secondly the wind usually shifts to come from some direction to the right of the wind direction before the front arrived (ie the wind veers (NH), but backs (SH)).
- Thirdly there is a noticeable drop in temperature.

The not infrequent minor tornadoes that are experienced in Atlantic Europe are most often spawned along cold fronts. They do not usually lead to loss of life (unlike their much more violent counterparts in the United States) and the damage to roofs and garden sheds etc is quickly repaired so that, apart from local newspapers, TV and radio stations, little is heard of them. Occasionally, as recorded in Chapter 20, a bigger than normal tornado does make the national and international press.

A sudden cold front

It is a winter afternoon with cyclonic weather. The wind is fresh but not strong. Then all at once there is a great gust of wind and hail lashes the windows. You hear through the roaring wind an unmistakeable rumble of thunder. The window panes mist up as they do when suddenly cooled. The whole angry episode lasts for maybe a quarter of an hour and then, although the rain does not immediately stop, it all calms down and in an hour or two it clears – often to a starry night, and there are slippery roads and pavements by the morning. A sharp cold front has passed.

Do all cold fronts act like this? The answer is, no! We have just described the sharpest of cold fronts and many cold fronts are quite meek and mild. They will have echoes of a more virile past but there will rarely be any thunder. The rain starts more or less suddenly, and is a mixture of continuous rain accompanied by showers at first. It then, as is in keeping with the behaviour of cold fronts, tails away with time.

However, these weaker cold fronts often do bring cooler, clearer conditions with possibly more showers to come and, even if no showers, heap clouds populate the sky – if not today, then tomorrow.

What about tomorrow?

When the fronts of a depression pass, the cold front is the last to cross us, so when you see the back of a cold front, it is very likely that tomorrow will at least start with a ridge of high pressure. There may be another low and its fronts to come but for a while the sky will be fair and populated with cumulus clouds.

When a cold front passes at sunset we have the conditions that have spawned the 'red sky at night' jingle. But not always – we will see why this saying works in Chapter 17 when we talk about colour in the sky. Just for now we'll note that the conditions when a cold front passes at sunset are most likely to lead to a night that will delight shepherds. There has to be some truth in weather lore that stands the test of time.

Occluded fronts

Occlusions are the result of an amalgamation between a warm front and a cold front. It is a meteorological fact that cold fronts travel faster than warm ones, so from the moment a depression is born, its cold front is going to be chasing and overtaking its warm front.

When the cold front begins to overtake the warm front ahead it drives in under the latter and lifts the warm air off the ground entirely. Gradually this process progresses all the way down the warm front and the two fronts together make an *occlusion*.

In Atlantic Europe, occlusions are the rule rather than the exception. Depressions that form many hundreds of miles offshore have plenty of time to occlude and therefore many of the fronts experienced over lands that look westwards into a great ocean (and this includes the western seaboard of North America) are occlusions. There are pure warm and cold fronts as well but often the front that is raining on you is an occluded one.

Photo 22 When a thundery outbreak threatens, the sky takes on some very strange-looking aspects. Beneath the cirrus and alto clouds a dark skein traverses the sky. It is an omen, but thunder will not break out yet.

Occlusion weather

When an occlusion approaches you often cannot tell that it is not a simple warm front. The cloud build-up is the same (photo 55). The rain starts in much the same way but there will not be sudden cessation as there is when a warm front passes. Also, as the front is old the rain may be more intermittent.

You may suspect that it is an occlusion when the steady rain becomes mixed with showery bursts. It is rare for that to happen with plain warm fronts. Then the weather should improve in the way it does behind cold fronts – gradually diminishing rain and then breaking into blue skies as the great mass of cloud moves away eastwards (photo 50).

It is best to see if an occlusion threatens from a weather map because if the front that is building is a warm one there is the risk of fog or at least poor visibility when it clears. That is most unlikely when an occlusion passes as the air to come is going to be cool and is most likely to create heap clouds – the enemies of fog.

Occlusions can drift about in the pressure systems and bring long hours of light or moderate rain. Sometimes they will get themselves more or less parallel to the isobars and then they are difficult to move. In these circumstances they may take on a serpentine motion – they are said to *wave* – and periods of more intense rain will occur sometimes accompanied by thunder.

In winter it is often old occlusions that lead to unexpectedly heavy falls of snow. Warm fronts do not normally snow while cold fronts are often moving quickly enough for the snow they produce not to be too disruptive.

8 showery weather

This jingle is about weather when heavy showers develop during the day:

Mountains in the morning
Fountains in the evening

Shower clouds have to be deep because their rain has to start as snow. Hence the mountains. The evening 'fountains' are a bit more problematic to explain because the time when showers are usually at their most intense is mid-afternoon.

The change in temperature of the land during the day can build heap clouds that are quite humble cumulus in the morning forming into towering cumulonimbus by the afternoon. The fountains from these can, and do, go on into the evening but usually they lose their potency by late afternoon or early evening. Evening is a time for the skies of heap-cloud days to clear and there is often a fine night (photo 7).

How do showers develop?

Showers occur in air masses that are relatively cool or cold and the visibility is excellent. It may become quite warm in the mornings of days when showers are going to erupt but the environment in which the shower clouds will develop does not know about this. It is, from their point of view, beautifully cold higher up and that is what showers need.

Such air masses have, very often, come directly from polar seas. They have warmed up somewhat at the bottom but for most of their height they are cold – and moist. They are ripe to build showers.

As the sun heats up the ground in the morning, local hot spots develop and the air over these rises as thermals. You cannot see thermals but you can visualise them as having the shape of enormous jellyfish pulsing off from the ground to find cooler regions higher up. The analogy also works if you imagine that they have a wake trailing back below them resembling the tentacles of a jellyfish.

As a thermal rises, it expands and cools so that at an altitude that is typically 2,000ft (600m), the moisture it contains begins to condense into cloud. The thermal current goes on rising until it meets a layer of air whose temperature is the same as its own and then it stops rising. Again, typically, this means the tops are about as high again as the base is from the ground, ie about 4,000ft (1km). These rounded lumps of cloud are cumulus and, if they do not grow any bigger, they are the typical clouds of fair weather.

The natural state of the atmosphere is for temperature to fall as you ascend so if a layer exists at some altitude where the temperature actually rises with height this is called an *inversion of temperature* or just an *inversion*. Inversions occur for several reasons which we won't go into here but just note that the reason cumulus clouds stop rising is that they have met an inversion.

Photo 23 There is no doubt that this is a hefty shower. The cloud tops, seen through the outriding cloud, are big and bulbous, and in the right-hand corner the horizon is obscured by heavy rain.

In the early morning it is normal for there to be an inversion just a few thousand feet off the ground. As the morning sun makes the ground temperature rise, thermals will start off from it and can sometimes punch through an inversion. The next time they are forced to stop rising may be tens of thousands of feet higher up and so their tops get into the cold realms where snow can form. Now moderate or heavy rain is the result and showers are born (photo 23).

Where do showers develop?

We have to face it – no one can ever say exactly where and when a shower will erupt, so it is on days when showers are forecast that you have to make up your own mind about the right-now weather. However, it always helps to understand what is going on.

The showery airstreams that invade Atlantic Europe are usually drawn southwards round the semi-permanent Icelandic low. They start off cold and moist and find themselves over increasingly warm water. This warm water acts to form a vast field of thermals and immense fleets of showers develop to cascade their snow, sleet, rain and hail into the watery wastes of the Atlantic.

However, here is an odd thing. By producing all these showers the atmosphere effectively kills them. The reason for this is that whenever clouds and rain etc are produced, a great deal of heat is generated at the same time. You only have to boil a kettle to realise that a sizeable (and expensive) amount of heat is needed to turn water into vapour. Conversely, when vapour turns back into water (as it does when clouds etc form), heat is released. The amount released in all these Atlantic showers is immense and what are they doing? *Taking heat from the sea and pumping it up into higher reaches of the atmosphere.* (See latent heat in Chapter 21).The eventual result is that all this heat being taken aloft produces an inversion, which stops the shower clouds getting too big. Thus often the showers that come ashore on the western coasts of Europe are not as virile as you might expect.

In showery airstreams, people living on the west-facing coasts may find their showers go on day and night because the sea does not change temperature daily (although it does on a monthly basis). If they don't get any showers or only weak ones, then an inversion is acting to cut the tops of the clouds down.

Over land, showers may not be about early but we have to allow for them intensifying during the day. Rising ground will set off showers that might not otherwise occur (photo 24). So wherever there is high ground not far from the sea, as there is, for instance, along the west coasts of Britain, Norway, parts of France, Portugal and Spain, clear, sunny skies over the coastlines will often turn into big clouds and showers as the air gets transported up the slopes. This also applies to the west coast of North America, where mountain chains stretch all the way from Alaska to California.

Photo 24 When high ground exists inland, onshore winds will often induce big heap clouds like these, which may not produce showers near the coast but may well do over the higher ground inland.

Where are the showers?

They have promised us showers, big ones, possibly with thunder, yet the morning has been very pleasant. Have they got it wrong or are we just lucky?

It may be neither, because around lunchtime, or just after, the blue sky and small heap clouds of the morning give way to dark forbidding shower clouds from windward. For an hour or two there are showers – some of them heavy – and you may hear an occasional clap of thunder. Then it all passes to leave a fine evening.

When this happens the culprit is usually what is called an *air-mass trough* (photo 21). Often showers tend to act like this – clumping together into lines across the wind that may look like a cold front. They are not fronts, however, because the air on both sides of the trough is the same kind, while on either side of warm, cold and occluded fronts it is different. These troughs seem to gather most of the clouds to themselves so that the hours before their onset are unnaturally sunny and it is often the same after they have passed.

Air-mass troughs tend to be most prevalent during the middle of the day and into the afternoon. This is the time when the ground is warmest and so is ripe for thermals to rise and develop into showers. The typical wind direction is NW – the direction of the cool northern seas.

Photo 25 Under this shower cloud, snow is falling. The particles that form the cloud are falling out of it, so soon there will be little cloud left and the snow will be over.

It may not always be a sunny morning, but while the heap clouds are often bigger than normal they do not develop into showers. Yet the forecast was pretty positive about the risk of showers, and if the forecasters, with their sophisticated methods, thought there should be showers, then you should too. Be suspicious – if the morning seems to be too good, is there an air-mass trough lurking somewhere over the windward horizon.

Another time when forecast showers do not develop is when the pressure is rising so that sinking air from aloft induces an inversion at altitude and stops any possible shower clouds growing up to high levels.

Am I in a shower shadow?

It is worth remembering that clouds can only hold so much water vapour. Lose much of it on a hill or mountain range and the area in the lee will be drier than normal. Such protected areas are called *shower shadows*. If you live in a shower shadow, you will get less precipitation of all sorts. There will be less rain when fronts pass and less snow in winter when the wind is favourable.

Of course, the best way to keep drier is to move inland. However, you do not have to. Near Colchester in northeast Essex there is a little place called St Osyth. Now, much of Essex is flat and in any case St Osyth is not far from the North Sea but even so St Osyth prides itself on being the driest place in England. This is because there is all of England and Wales to the west where most of the rain comes from. There are the North Downs and the South Downs to the southwest – a particularly rainy quarter. There is the great bulge of East Anglia to the north that robs northerlies of their moisture. From every direction except the east St Osyth is protected – and the dry Continent is to the east.

You can probably think of areas near you that seem to bear charmed lives when it comes to weather and maybe a contour map will throw up some higher ground in the direction of rainy quarters that can account for this.

Do you get more than your fair share of rain?

Just as there are parts that are unusually dry, so there are those that are unnaturally wet. Of the areas most likely to be wet, we can cite the hill farms that overlook the sea on west-facing coasts. As it takes time for rain clouds to form as they ascend hills and mountains, so the coastal belt itself can be quite dry, but move a few miles inland and rain and drizzle are the norm. It is worth remembering that rivers start from high mountain ridges that form watersheds. It is bound to be wettest where it's highest – as well as being snowiest.

However the coasts of the Netherlands, and northwest Germany, are low with very few hills. Yet, when the wind (in the winter half of the year) turns into the northwest, these areas are often swept by showers of rain or snow because they stick out into the North Sea. However the air does not have to get very far inland before the precipitation dies out. Now, by human standards, the North Sea is cold but by weather standards it is often warmer than the air that blows over it. This induces showers to develop over the sea and blow in across the coastal areas, making these regions wetter or more prone to snow showers that may not survive very far inland.

On showery days, how can sunshine mean more rain?

You are having a succession of showers and then you get a break. The sun comes out – it's good news – or is it?

As we saw above, it is thermal currents from the ground that initiate heap clouds. There is little chance of thermals developing under grey skies but as soon as the earth sees the sun it begins to have the potential to form them. The air is already stuffed with moisture from the recent showers and so is ripe to build new shower clouds. So enjoy the sun while it is there – but don't expect it to last long.

The same applies when cold fronts and occlusions pass. The high cloud edge of a retreating cold front or occlusion is like a shutter that opens to reveal the sun (photo 19). There are usually clear skies for a while when a front clears but this is only until the earth has absorbed enough heat to set off thermals. Then the heap clouds begin to grow and often grow big enough to produce showers. The clouds may be ragged and grey at first because there is so much moisture about.

Coastal dawn

Asleep in your bunk on a yacht or maybe camping or caravanning near to the coast you are surprised to hear the sound of showery rain, which you did not think looked likely when you turned in. The sun may be just up as you peer out (or maybe scamper out to rescue things you left out overnight). Where has this come from?

The answer lies in the fact that coastal sea is usually warmer than the land at this time of day, so when cool nocturnal winds sidle off the coast overnight the warmer sea erupts this cool air into shower clouds. Often these local cumulonimbus clouds drift inland in the early morning, so disturbing an otherwise dry night. Usually they will all be gone after breakfast.

Why is it colder after rain?

I know that when you are wet you feel colder, but I am referring to how the air feels. To evaporate water you need to give it heat, so when the ground begins to dry up after a shower it gains the heat it needs from the ground and the air. This makes the air cooler.

Other kinds of showers

What I call real showers come from big, blustery, often individual heap clouds usually with some sunshine between. However, any rain that starts and stops within a short period will be described in forecasts as showers.

Often these come from layers of medium-level cloud. The layers will be altostratus but the showers come from showery areas embedded in them. Sometimes these may be thundery in nature with big raindrops and perhaps some rolls of thunder. If it is dark enough you may well see sheet lightning with these kinds of showers (photo 22).

Forecasts for the general public cannot be too technical, so while the forecast could describe the showers more adequately, they have to use just two descriptions. It will either rain, in which case the precipitation will last for a long time – maybe hours; or there will be showers which will embrace all kinds of rain that starts and stops. The rain periods may last an hour or so and still be called showers.

Look to the sky

If rain starts gently and then gradually increases you have a warm front or an occlusion. If it starts heavy (or even moderate) you have a cold front or a showery trough or just simply showers. You can usually sort out which from the look of the sky.

wind

Every wind direction has its weather.

So says a well-known bit of weather lore.

North winds tend to be cold for the time of year and bring showers to places exposed to them. They are traditionally associated with snow in winter.

Northeast winds are dry and cool and may be bitter in winter but benign in summer. On the east coast of England, in the summer half of the year, they can produce coastal fogs called 'haars' – maybe so-called because they are as thick as the hair on your head.

Easterlies have a reputation for being capricious with strange wind and weather patterns. They occur when pressure is low to the south and may presage thunder.

Southeasterlies can be dry and warm for the time of year when they come from the direction of the Mediterranean but, nearer to the UK, they also blow ahead of coming depressions under skies laced with cirrus clouds.

Southerlies are again like southeasterlies when bad weather is on the way. If they are from some sea area far to the south they may be very warm in summer and mild in winter. They are very often cloudy.

Southwesterlies are typically cloudy, drizzly and generally overcast. They bring hill and coastal fog and are usually mild for the time of year. However, not all south-westerlies are so wet – you can get 'dry' ones which bring balmy weather.

Westerlies tend to bring us some of our best weather with fleets of modest cumulus clouds and normal temperatures. When they follow the fronts of depressions, they can produce showers.

Northwesterlies are typified by showers and well-washed skies. They are often cool and in coastal places showers may occur both by day and by night.

These descriptions refer to the large-scale winds – the winds that come with depressions and anticyclones – and there are going to be many times when your wind is due to local influences such as small depressions nearby, the hills and the valleys where you live or coastal winds if you dwell near the sea.

Unnatural weather

There are times when the wind is south to southwest – a direction that is associated with warmth, humidity and layer clouds but not with blustery showers. But big shower clouds are blowing in just as they might on a northwest wind. What's wrong? The answer is usually that there is a low-pressure centre not far out to the west bringing cool, unstable air around it and so while the wind is from the south, the air has originally come from the northwest or north.

What causes the wind direction?

The short answer is the shape of the local pressure pattern. It is a well-known fact that the lines of equal pressure (isobars) that form a weather map also show you the way the wind is blowing, both in direction and speed (Fig 6.1).

However, local effects can alter the isobaric wind. Winds tend to follow valleys and skirt round hills and mountains. They will try to blow along water channels and along lakes rather than across them and, when they are light to moderate and the sun shines, they abandon the isobars and blow directly from sea to land as seabreezes. Light winds overnight lead to land breezes that sidle off the coast onto the water during the early hours. These are aided by air sinking from high ground just inland from the coast. If the coast is mountainous these nocturnal winds can get strong.

In mountainous districts, with good weather, valley winds blow towards the mountains in the afternoons and mountain winds blow back down the valleys during the night.

Why doesn't the wind blow directly from high to low pressure?

The short answer is because of the effect of the Earth's rotation. This deflects winds to the right of their path in the northern hemisphere and they end up blowing to keep low pressure on their left. Go to the southern hemisphere and they are deflected to the left.

Because of this effect we get a useful rule:

*Stand with your back to the wind and pressure is **L**ow on your **L**eft hand.*

This is easy to remember because of the L in both low and left. In the southern hemisphere, just face the wind and the rule still applies. We see this in practice in Figs 6.1 and 7.1.

What about wind speed?

It was Admiral Sir Francis Beaufort who first attempted to give mariners an idea of the wind speed by relating it to the amount of sail that a man-o-war could carry and the speed she could make. Now, today, 200 years later, when mariner and landlubber alike need other criteria to assess the strength of the wind, we still refer to it as the Beaufort Scale of Wind Force.

The unit of a nautical mile per hour – the knot – is used internationally for wind speed and is also useful for those who use the metric system because a metre per second is just about 2 knots. If you want it in miles per hour then just add a bit to the speed in knots. For example 20 knots = 23 mph and even 50 knots is only about 58 mph. Usually you won't need anything more accurate than this.

Photo 26 In Beaufort force 1 winds, smoke just drifts.

THE BEAUFORT SCALE OF WIND FORCE		
FORCE	DESCRIPTION	APPEARANCE
0	Calm	Smoke rises vertically. Sea mirror-smooth.
1	Light air	Smoke drifts. Wind vanes do not respond. Scaly ripples but no foam crests.
2	Light breeze	Wind felt on face. Vanes respond. Short wavelets that do not break.
3	Gentle breeze	Light flags extend. Leaves in constant motion. Large wavelets. A few scattered white horses.
4	Moderate breeze	Flags fully extended. Paper and dust may be raised. Fairly frequent white horses.
5	Fresh breeze	Tops of tall trees in noticeable motion. Many white horses.
6	Strong breeze	Wires whistle. Large waves form. Some blown spray.
7	Near gale (moderate gale)	Whole trees in motion. Foam blown in streaks.
8	Gale (fresh gale)	Twigs broken off, walking against wind impeded. Moderately high waves. Well-marked foam streaks.
9	Strong gale (strong gale)	Slates removed, fences blown down. High waves. Crests begin to topple.
10	Storm (whole gale)	Very rare inland. Trees uprooted. Considerable structural damage. Very high waves. Whole sea surface appears white.
11	Violent storm	Exceptionally high waves. Visibility impaired by flying foam and froth.

Why do gales happen?

Winds blow at all kinds of speeds from flat calm to something like 150 knots (170 mph or 75 metres per second), which is the highest recorded wind speed in the United Kingdom. However, that was a record set in the high, exposed Cairngorm Mountains in Scotland in 1986. Higher wind speeds occur in tornadoes in the United States or the typhoons of the China Sea but not usually in the weaker tornadoes of Europe.

On the whole, the strongest winds we experience appear around depressions. These may rise to 50 knots, which is described as *storm force*, but more normally when gales are forecast, the wind on average is not expected to rise to higher than 40 knots. However, this is at sea. Inland if a gale reaches 40 knots it is a rare occurrence and considerable structural damage will occur.

Gales often appear when a depression deepens near an anticyclone that is loath to leave. For instance, in Fig 6.2 if high Y does not give way but the central pressure of low W tumbles, more isobars have to cram into the space between the two.

When isobars get closer together the wind that blows between them must have increased. We can think of the isobars as akin to contours on a topographic map. A set of closely spaced contours indicate a steep slope or gradient of pressure. Roll a ball down a gradient and the steeper the gradient, the faster the ball rolls. By analogy a closely spaced set of isobars indicates a steep *pressure gradient*. However, where the ball analogy breaks down is because of the effect of the spinning Earth. If our ball now represents the wind speed it has to be directed along the contours like a motorcycle rider on the wall-of-death. Even so, just as the wall-of-death rider has to travel faster to stay upon the steeper parts of the bowl on which he's riding, so the steeper the pressure gradient, the higher the wind speed.

So all over a weather map we can assess the relative wind speeds by judging how far apart the isobars are. For example, in Fig 6.1 it will be calm between the Azores and Portugal. Over southern Greenland the wind will be southerly and severe gale force. There is very little wind over Scandinavia and light winds under leaden skies over southern Italy. Britain has moderate winds from the north, which will be cold as they have come from northern seas.

Photo 27 Beaufort force 7–8 – the cloud appears ragged because it is being formed by turbulent over-turnings in the strong wind.

How hard will it blow?

Frankly this is difficult to assess so the forecasts are essential. The modern forecasting revolution, where extremely powerful computers forecast what the isobars will look like days in advance, means that rarely will there be strong winds without the forecasters knowing about it and issuing warnings. Together with the input of human forecasters, the one thing you can certainly rely on is good general forecasts for depression weather. The computers handle such situations very well – they may not be quite so good when anticyclones settle – and there are sometimes small local depressions that alter the weather in a local area out of all recognition. It is these that the computers may not fully recognise. Great efforts are being made to forecast these smaller-scale features with computers dedicated to their recognition, movement and severity.

Photo 28 Beaufort force 6–7 – a strong gust explodes the gulls into the air.

Still, there are some useful signs for the layman – some of which have been in use for centuries.

> *When the rain's before the wind, then your halyards you must mind.*
> *But when the wind's before the rain, soon you may make sail again.*

What is behind this profound bit of weather wisdom? It is to do with the speed of the upper winds – the ones that move the cirrus clouds across the sky. It is well known that when the high-level winds blow very strongly – as they do in jet streams – then the depression associated with jet winds is also likely to be vigorous (photo 33).

HIGH-SPEED JETS

Jets can blow at 100–200 knots and when they do they carry the weather forward at high speed so that it outruns the low that spawned it. It is in these cases that rain can break out well ahead of where you'd expect it. At this time the surface wind has not increased very much but the early rain says that the wind, when it comes, will be gale or even severe gale force.

On the other hand if the wind gathers early, before the rain has started, it is likely to pass relatively quickly.

Why are winds sometimes gusty and sometimes not?

The wind you feel when out walking etc is not confined to the first few metres (tens of feet). Because it blows over surface objects like houses, factories, woods and copses it develops big eddies. These eddies get bigger as the wind speed increases. They are aided by up-and-down convection currents during the day and in this way large parcels of air from higher up get mixed in with the surface wind.

The wind higher up is blowing faster than the surface wind, which is being slowed down by collision with objects on the ground. So the eddies bring faster wind from above. When one of these parcels of faster wind arrives it forms a gust that will be followed by a lull. Lulls are the true surface wind and when the latter is invaded by these chunks of faster wind, a succession of gusts and lulls develops. This chunkiness is how the real wind blows. The typical time for gust-lull changes is minutes. We see the effect of a gust in photo 28.

Why are there big gusts around showers and thunderstorms?

The great cumulonimbus clouds that generate showers and thunderstorms often produce shafts of very intense rain and hail. The shafts bring down the wind from higher up and spread it out around the storm as big gusts. These gusts come from the direction of the shower or the storm and may temporarily double the wind speed. At the rear of these big shower clouds the wind is most often light. When you see the rain curtains of an approaching heavy shower or thunderstorm (as in photo 16), allow also for the gusty winds that will accompany it.

What's bad about easterlies?

Easterly winds have a bad reputation. Izaak Walton in his *Compleat Angler* says under 'Signs of Rain':

> *When the wind is in the north,*
> *The skillful fisher goes not forth;*
> *When the wind is in the east,*
> *'Tis good for neither man nor beast;*
> *When the wind is in the south,*
> *It blows the flies in the fish's mouth;*
> *But when the wind is in the west,*
> *There it is the very best.*

This stricture against the easterly wind has applied for millennia and to places as distant from Atlantic Europe as the Eastern Mediterranean, as is shown from quotations from the Old Testament. In Jonah 4:8 we have:

God prepared a vehement east wind

And in the Book of Ezekiel 27:26

The east wind hath broken thee in the midst of the seas

If you study the way the east wind blows within itself you will see that it does very strange things. The west wind blows turbulent and gusty in the way you instinctively know is normal. Not so the easterly. It often goes to sleep for minutes on end and then climbs out of its somnambulism to blow for as long again before seemingly expending its limited energy and sinking back into sleep again. At the same time its direction waves wildly around, sometimes blowing from directions that are 90 or more degrees apart. Another thing about the easterly is that it is often an unnaturally warm and humid wind.

Always treat the easterly wind with respect – you never know when it will spring a surprise on you, especially at night.

Are night winds different from day winds?

The general answer is yes. Over the land, on almost all nights, the wind blows more gently than during the day. This change happens over the late afternoon and evening. However, in coastal districts the wind over the sea increases in speed with the onset of night, especially when the wind direction is similar to the lie of the coast. The difference is greatest with winds that are blowing more or less parallel to the coastline. Here the wind just off the coast may be twice its value ashore. In most cases the wind at night blows more smoothly than it did during the day as there is no convection.

Where hills or mountain foothills come close to the shoreline the night wind may rise to near gale force from the land, especially when the mountain tops are snow-capped.

What is turbulence?

Take time to look at the antics of a weather vane. Assuming it hasn't got stuck, it will never be still – swinging backwards and forwards over what are usually small angles. The typical time for these changes is in seconds.

These smallest of changes made by the wind are due to turbulence, and the little eddies induced in the wind as it blows over the Earth's surface (photo 27). So there will be more turbulence over the land than over the sea and the more objects there are impeding the wind, the greater the turbulence, and this is why a weathervane is normally in constant motion. However, there are rare times when it will remain apparently fixed in one direction for minutes on end, despite there being a wind you can certainly feel.

Why is the wind blowing so smoothly?

Sometimes, and this happens most readily in the evening and overnight, the wind goes into a very smooth mode with hardly any variations at all. This coincides with the establishment of an overnight inversion not far from the ground. The surface wind cannot get any help from faster wind above the inversion and so loses momentum. When it blows over flat surfaces like the sea, big lakes or mudflats, then it can blow smoothly without the turbulent variations we get earlier in the day.

What is a seabreeze?

One of the delights of the beach on hot summer days is that there always seems to be a cooling breeze (photo below). This breeze comes from some direction seaward and it is cool because, compared to the land, the sea is cool.

Whenever the sun is out over the coast in spring and summer there is potential for a seabreeze. The sun warms the coastal land and thus the air above it. The heated air expands and this leads to some of the air over the land sidling out over the sea at altitude. Thus air shifts from the land to the sea and so pressure over the land falls while that over the sea rises. Near the surface, sea air now moves in to try to adjust the pressure difference. This is a seabreeze.

Seabreezes blow at 10–15 knots on most coasts of Atlantic Europe but they can become strong in the Mediterranean and similar climes where they can make 25 knots or more.

Photo 29 The perfect beach morning – even the cirrus cloud above the fair-weather cumulus does not indicate any great change. This is the kind of morning – sunshine, light winds, cumulus clouds – when seabreezes are inevitably going to blow. These onshore winds have a life-saving role in that they will blow inflatables of all kinds ashore.

The marine layer

At Longbeach, California the coastal strip is backed by a very hot desert. There ought to be a massive seabreeze but there is not. The seabreeze is regular but muted and it is because of what is called the *marine layer*. This is an inversion layer that puts a lid on the seabreeze and will not allow its air to escape upwards. Thus the breeze cannot blow at all strongly. This happens elsewhere also. The hottest days rarely see a strong seabreeze because of the subsidence inversion that inhibits the thermals that seabreezes need.

What is a landbreeze?

On fine-weather evenings it often falls calm or there is just a very light breeze. In these conditions, under clear skies, the land cools but the sea does not. Now the cooling air over the land packs down, setting up the conditions for air to gently move bodily from sea to land not far above the surface. In direct contrast to a seabreeze, air from the coastal waters, say a thousand feet (300m) up, sidles gently in over the land and the result is that pressure rises over the land and falls over the sea. Surface air gently moves from land to sea and we have a landbreeze. Landbreezes may not get deeper than a few tens of feet (maybe 10m) and they are greatly impeded by buildings, copses etc near the coast.

Landbreezes usually don't amount to much – just a few knots at most.

What is a nocturnal wind?

Landbreezes would not normally be even moderate in strength were it not for another effect that helps them on their way.

Katabatic winds are winds that blow downhill when the slopes cool in the evening and night. Near the coastline a landbreeze only occurs if the land cools at night and a katabatic wind only occurs for the same reason. Together they form one, usually light, breeze that blows from land to sea. This is called a *nocturnal wind*.

Nocturnal winds are aided by lines of coastal hills eg the South Downs on the Sussex coast of England or the coastal high ground of Artois, Picardy and Caux on the northwest coast of France. In some places, such as the coastlands near the Franco-Spanish border where the mountains of the Pyrenees are not far away inland, the nocturnal wind becomes something to be reckoned with. It may even get to gale force when other conditions are favourable. Similar strong downslope winds exist along the east coasts of the Adriatic Sea and in fact anywhere that has mountains not far inland.

Why all the sunshine?

Coastal resorts pride themselves on how much sunshine they record. Particularly favoured stretches of coast become known as rivieras. In England, for example, there is the South Coast Riviera that stretches from Bournemouth in Dorset eastwards to Kent. Here the total hours of sunshine in the spring and summer are truly remarkable.

It is no coincidence that these are also the coasts that enjoy the most seabreezes and you don't have to look far for the answer. When a breeze starts (which it usually does in the mornings of otherwise clear, quiet, sunny days), it often does so against a gentle wind from somewhere inland. Here we have an impossible situation – two winds blowing towards one another. Their only chance of escape is upwards and a seabreeze front forms more or less parallel to the coast, which acts like a chimney, up which the converging winds can escape.

The air that consequently goes aloft has to come down over the seaward side of the front and so, as sinking air warms up, clouds over the coasts tend to evaporate away leaving sunny skies. Journey inland and it will normally be much more cloudy. So the coastal resorts are not telling fibs – they really do get more sunshine because of the seabreeze regime.

Photo 30 In Poole Harbour, looking towards Brownsea Island, the open sea is to the left. A seabreeze is just starting to blow. There is little cloud on the seaward side but a distinct line of it on the landward (right).

Extreme winds

Occasionally – luckily very occasionally – prodigious winds blow up. Force 13 on the Beaufort Scale is hurricane force, defined as 63 knots or more, so when 80-knot-plus winds ravaged northwest France and southeast England on the night of Thursday/ Friday 15/16 October 1987 it was a once-in-a-lifetime occurrence.

This unprecedented tempest came largely from the south or southwest depending on where you were, and almost unbelievable wind speeds were measured. For example, at Dungeness in Kent they had 90 knots sustained wind for a full three hours and on the south-west side of the Brest Peninsula the wind was recorded at 119 knots.

The wind came to its maximum in southeast England just before daybreak when very few people were about and so there were, mercifully, few casualties. Not so for the electricity pylons and poles – some people were not reconnected for three weeks. However, what many people wept over most was the loss of the trees. Millions were felled and leafy places they knew and loved were totally laid waste. Mature trees were just snapped like matchsticks. No less than 80 per cent of the trees in the Tunstal and Rendlesham Forests in Suffolk were felled and those of us who had previously visited regularly could hardly believe the degree of devastation inflicted by this extraordinary storm.

Photo 31 A false-colour shot of Hurricane Andrew making landfall on 24 August 1992. This Category 4 hurricane produced gusts up to 150 knots, caused a 17-foot storm tide and damage costing $26 billion. Photo courtesy of NOAA.

10 will there be bad weather?

When we think there may be bad weather in the offing, it presupposes that at the moment the weather is good or at least reasonable.

The weather map, before a depression spreads its clouds and rain across us, normally shows a ridge of high pressure. The typical sky is like photo 6 – but what is the streaky cloud way up above the fair-weather cloud?

Photo 32 In the evenings of quiet days, the heap clouds of the day spread out but there are still considerable gaps between the islands of stratocumulus created. The lower air has become polluted with dust, fumes etc and the sun shining through the gaps into this murky lower air creates the so-called 'crepuscular rays', which indicate a fair night.

We have already seen in Chapter 5 about how, when a warm front or occluded front is coming, it is the high clouds you need to watch.

Signs of bad weather

As well as cirrus-watching, a good sign of probable deterioration is when the individual cumulus clouds flatten and spread into a layer, called *stratocumulus* (see photo 32).

Otherwise low clouds disperse as the cirrus above thickens and partially begins to cut off the sun. Although we have already looked at the stages in the cloud build-up before a front brings rain (or snow), it is so important that we are going to expand on it here.

CLOUDS TO LOOK FOR WHEN BAD WEATHER IS APPROACHING

The first cloud to appear is the high, white, teased-out filaments, hooks and banners of *cirrus*. Cirrus may take an hour or several hours to build up (photos 14, 20, 33, 34 and 38). Then you often become aware that the sun (or the moon) has a ring halo around it. There is a milky veil across the sky called *cirrostratus*. The halo may not be visible for as long as the cirrus (photo 15).

Photo 33 Here is a sky you should not ignore. It is a form of very high cloud that runs ahead of bad weather. It is called jet-stream cirrus and is spawned by intense depressions. If you see these banners, which appear to converge towards the horizon, check the forecasts for gales and bad weather generally.

The ice clouds above are very high, possibly as high as 5 or 6 miles, but then darker cloud banks start to come in at a lower level. These are *altostratus* and they are darker because they are composed of water droplets (photos 18 and 35). They may soon cover the whole sky – the ice clouds are still up there but you cannot see them. Rain is much closer and much more certain now.

As the altostratus layers deepen, the sun will disappear more or less slowly into the murk and eventually it will begin to rain. Now the clouds have become deep enough for snow to appear in their tops and rain is the result. These deep rain clouds are called *nimbostratus*, a name that means rain-bearing layer cloud. Photos 12 and 17 show this kind of cloud.

During this time it is normal for the wind to back into the south or southeast and to increase in strength. When it doesn't, things are abnormal and maybe you need to find out what the professionals think about the situation.

WATCH THE CIRRUS

It may appear thin and wispy but cirrus is the most important cloud in the sky when it comes to foretelling deteriorating weather. The winds at cirrus levels are often going to be very strong when bad weather is in the offing. High-speed rivers of high altitude winds (*jet streams*) blow in association with bad-weather systems. These winds may get as strong as 200 knots although 100–150 knots is more common.

Photo 34 Another sure sign of strong wind to come is 'hooked' cirrus, seen here in the centre of the picture. The hook is formed by a streak of ice crystals falling from the head in the high sky as an ice shower and suddenly bent backwards by very strong upper winds – which mean strong surface winds are on their way.

Further, these winds are never in the same direction as the surface winds when bad weather is imminent. For example, the jet winds will blow from the west when the surface wind is from the south. This is a typical set-up when bad weather is coming in. Conversely, as the weather eventually clears, the typical surface wind is north-westerly. If we see cirrus clouds moving from the southwest behind a clearing cold front, then you can expect better weather tomorrow.

Because it is some 6 miles (10 km) aloft, you may find it difficult to detect motion in bits of cirrus – but when you can, the jet streams are blowing at 100 knots or more. The kind of sky that goes with coming bad weather often looks like photo 33 (although this is an unusually fine example) and skies that have this banner-like look, together with the clue of the excessive speed of the cirrus elements, indicate that you are in for trouble. In these circumstances surface winds may rise to force 8 over land and force 10 at sea.

A common scenario is when the surface wind has backed into the south and the cumulus has disappeared, leaving a clear vault so you can easily see the cirrus banners streaming in from the west. However, the crossed directions of high altitude wind and low-level wind do not have to be west and south. They can be, say, north and west (possibly snowy situations) or south and east, most often experienced when thunderstorms are due. In the latter situation the upper clouds do not need to be as high as cirrus. They can be the kind of alto clouds we see in photos 40 and 41.

What about occlusions?

The sequence of cloud types above is the one that accompanies virile warm fronts, which are often part of equally virile depressions. However, the majority of the fronts that cross the coasts of Atlantic Europe are older occluded ones. Occlusions bring the same clouds and rain as warm fronts but often the weather passes more quickly and there may be holes in the clouds through which patches of blue sky can be seen.

How long before the weather really deteriorates?

Long foretold – long hold
Short forecast – soon past

This is a truism that reflects life. Courtships that take time to develop have a much greater chance of producing a stable relationship than the whirlwind romance. The same goes for weather. When the bad weather clouds come in taking their time, then the poor weather is also likely to last a long time. However, if the cloud types all appear in the sky more or less at once (as they do in photo 20), then the rain to come is equally likely to be short-lived. The first line above refers to young active warm

Photo 35 When a warm front or occlusion is approaching, the solar halo disappears as darker layers of cloud invade the sky. Now the cloud is thick enough to begin to obscure the sun. This is a typical rainy sky – it is not raining yet, but it will.

fronts, while the second describes the short-period changes often associated with old occlusions.

What happens when a warm front passes?

Firstly the rain stops. Secondly there is a distinct break in the clouds – but don't trust it as it will often close in again. The temperature will rise and it will often feel very humid. The wind will often veer to the southwest.

It is usual with the wind from the southwest for the low clouds to close in again and it may well drizzle, especially on windward coasts. On hill and mountain slopes facing the wind it is very likely to rain or drizzle, but on low ground it is often dry (even if it feels damp in the wind).

Hilltops and especially mountaintops will develop cap clouds even when there are few clouds elsewhere and people hillwalking or climbing should recognise that the sequence of clouds that runs ahead of warm fronts and occlusions has a second message:

When the rain stops get your bearings in case of sudden fog.

What's a warm sector?

This is the more or less wedge-shaped region of warm, humid air that occupies the space between the passage of the warm front and the onset of the cold front (Fig 7.1).

Weather in warm sectors can be many things associated with warm, humid air from the subtropics. There can be fog (or poor visibility generally) over the sea and on coasts facing the wind (photo below).

There may be drizzle anywhere – but often there is not. It depends on how humid the air is. Cloudiness is the trademark of warm sectors. However, in summer the sun may burn off the cloud and leave a stiflingly hot day. When the fronts are occluded the warm sectors that appear between warm and cold fronts have been squeezed aloft.

What about the cold front?

The way depressions form out on the Atlantic presupposes that, some time after a warm front has passed, there has to be a cold front to follow – often several cold fronts one behind the other, each making it a bit colder.

Photo 36 A seagull cruises a steep coast bathed in sunshine, while not far to seaward is a blanket of sea fog. When coasts face the sun, they may well disperse fog close inshore; however, those who seek these inshore waters must be very careful to avoid rocks and shoals.

A cold front, by its very name, has colder air behind it, but when cold air tries to oust warmer air the result will often be some nasty weather for a time.

The rain from a warm front starts lightly and becomes moderate with time but it all tends to happen slowly. Not so with a cold front. Here the sudden arrival of cold air lifts up the warm air and sets off sometimes violent showers and thunderstorms at the same time as it rains heavily.

There are often nasty squalls and almost invariably a veer of wind (NH, back in the SH). Things are often cold, wet and chaotic when cold fronts appear – sometimes without much warning. But a cold front's weather is like a warm front's in reverse. The rain may start heavy but it soon tails away to moderate and then light before ceasing altogether. On the whole, the weather of a cold front passes twice as quickly as that of the warm front.

What's behind the cold front?

Eventually the sky lightens and breaks behind a great swathe of high cloud, including cirrus. Then, if it is daylight, there is a blue-sky gap before heap clouds begin to build. These may remain too small to produce showers but at other times heavy blustery showers of rain (or snow) set in. The air now feels different. It is cooler and clearer – you can often see for tens of miles (or kilometres) and you could now be in for a spell of better weather.

Sometimes this more benign spell will last for days and at other times only for hours when another depression is hard on the heels of the one just gone.

What's good about a red sky at night?

Red sky at night, shepherd's delight
Red sky in the morning, shepherd's warning

You get a red sky at sunset when the sun, now low on the horizon, can shine on high clouds.

As weather tends to travel from west to east, so these high clouds most probably are those of a cold front or occlusion that is passing. The evening is a time when the heap clouds of the day cannot form, so there is often little or no low cloud between you and the western horizon. That indicates a clear night to come and so shepherds and others are delighted (photo 37).

Spectacular sunsets are often provided by the clouds of older cold fronts (or occlusions), which have been fragmented and spun around the sky by sinking air from higher up (photo 8).

We cannot leave the red sky at night without mentioning the red sky in the morning (photo 55).

Photo 37 Nimbostratus, the deep-layer cloud of bad weather clears while underneath there are lumps of what is often called 'scud'. One of these lumps has caught a shaft from the setting sun. Scud is typical of clearing fronts.

Now the high clouds on which the rising sun can shine are those of a warm front (or occlusion), which implies a coming day of lowering cloud and rain. As winds usually increase with the onset of frontal weather, so the red sky in the morning presages a nasty day.

What does wind direction tell us?

In fair weather the wind is most often somewhere in the west but if you take careful note you may see that the wind direction is shifting towards the south. We say it is backing and a backing wind is often a sign of deteriorating weather. There is a saying:

A veering wind will clear the sky
A backing wind says storms are nigh

December 1999 – excessively stormy over Europe

Three violent storms hit Western Europe in December 1999. They claimed more than 130 lives and the total economic loss was estimated at 13 billion euros.

The first intense storm struck Denmark and the northernmost part of Germany on 3 December. Coastal winds gusted to nearly 100 knots (more than 50m/s) and the winds caused a storm surge that led to large areas of coastal flooding. In Denmark this storm was described as the most violent of the whole 20th century.

The weather went comparatively quiet for the next three weeks and then, on 26 December, another intense storm rushed across the Atlantic at the amazing speed of 75 knots to leave a trail of destruction from northwest France to southern Germany. Again the winds gusted to 100 knots. Notre Dame de Paris was damaged and, across the whole path of the storm, immense numbers of trees were uprooted. The gardens at Versailles were extensively ravaged by the wind.

Not content with this, the very next day, a third storm – further south than the day before – swept across northern Spain and southern France. The winds were not quite as strong but the excessively gusty wind still reached 80 knots in places, with many roads blocked by fallen trees and widespread loss of electric power that, in some cases, lasted for several days.

Barometers tumble at record rates when such winds are generated but in this case they were phenomenal. At Rouen, between 0300 and 0600 on the morning of 26 December, there was a fall in pressure three times greater than that usually associated with storm-force gales. When the pressure rose behind the depression it climbed at an even more staggering rate, approaching five times the normal storm values.

This was recognised as a very odd storm with sustained winds of hurricane force that came from many directions plus torrential rain, heavy thunderstorms and snow. The storm on the following day, which went further south, was almost its twin but it travelled at a mere 65 knots from roughly the point where the one of 26 December had started out on the Atlantic.

Winds that veer move in the same direction as the sun while shifts in the opposite direction are moving back against the sun. As bouts of worsening weather can come with all wind directions, we do not have to believe that it is only west winds that back ahead of deteriorations.

In a run of weather, always check the forecast when the wind backs because it is just as likely to be ahead of troughs of cold showers from the north or northeast as it is to be ahead of fronts from the west.

Winds will often back when thundery outbreaks are approaching. For example, in lands surrounding the English Channel there is an especially nasty thundery situation called a *Spanish plume outbreak*. Because Spain gets so hot during the late spring and summer, it sends a high plume of hot air northwards. This interacts with cooler air and sets off thunderstorms that are maybe two miles (3,200m) above your head. Before that happens the wind backs into the east and often blows moderate or even fresh.

Photo 38 When cirrus appears to fall down the sky like this, there is not likely to be much wind.

11 thundery weather

We've learnt a good deal about thunderstorms since the Second World War, but there is still the primeval fear of the unpredictability of lightning and where it will strike. No amount of scientific research can tell you if you will be struck as the flashes multiply and the thunder rolls around the heavens.

Perhaps what is most amazing is how few buildings get struck. If you are indoors, your chances of receiving any personal damage in a storm are minimal. Only in the open, and carrying metallic objects like golf clubs and umbrellas, do your chances of being struck really amount to anything.

What is a thunderstorm?

Basically a storm is a slow explosion of warm moist air, but there are different types of storm.

To make sense of thunderstorms you have to embrace a theory that first appeared in 1948 as a result of research carried out in the United States. It was then that they recognised that storms occur in individual units called *cells* and that, just like us, cells have a family history (see Fig 11.1).

You cannot have a storm smaller than a single cell and single-cell storms are most often associated with passing cold fronts, thundery troughs or just bigger-than-normal showers.

Sometimes vigorous cold fronts can spawn a whole line of storm cells. This can be very dangerous for light aircraft which, if they have to fly through the storms, can become iced-up and thus become potential candidates for fatal crashes.

However, what most people recognise as a thunderstorm is the weather condition that erupts on the afternoons of hot days and may last for hours. This is a multi-cell storm with cells producing thunder and lightning over a wide area and extended period of time.

Parents and daughters

A storm cell keeps going due to its updraughts. These are more or less vertical shafts of wind inside the storm cloud (Fig 11.1B).

The wind speed in the strongest updraughts can be as high as 60 mph (30m/s) or more. Such updraughts are rare but you can tell when the storm cell affecting you has such strong updraughts because the hailstones come down like a rain of icy golf balls – or bigger. Most of the time the updraughts just produce normal pea-sized hail but even then, in the worst scenario the hail may lay inches – even feet – deep.

A storm cell produces updraughts usually for about half an hour or less. This is when it is shooting lightning bolts about the sky. After that it loses its virility and just rains.

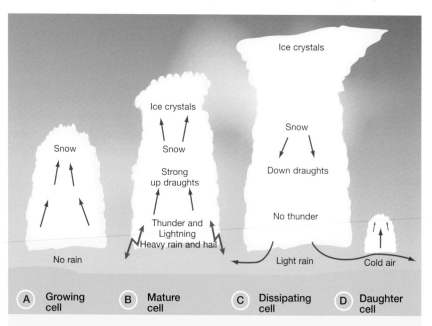

Fig 11.1 The cell theory of thunderstorms sees embryonic cells growing in unstable air so that in, typically, half an hour they are deep enough to have snow and ice in their tops. Strong up draughts build and heavy rain and hail occur, together with thunder and lightning. The cell is now mature. Soon after reaching the mature stage, down draughts set in which bring down whalebacks of cold air that spread around the storm. These lift the surrounding warm air and set it rising. This rising air may now induce the creation of a daughter cell that will grow beside its parent and take over as the parent dissipates. The parent cell will still make its presence felt by the anvil top it builds. Cells are most active when their cloud tops are round and knobbly.

So how do storms keep going for hours? The answer is by breeding daughter cells next door to them. The heavy rain and hail drags down cold air and this spreads around the storm. It dives under the warm air nearby and lifts off another incipient storm cell (Fig 11.1D).

Just like life, this daughter grows alongside its parent and, as the parent grows old and dissipates, the daughter takes over. There may be several generations this way and while we may speak about 'a thunderstorm' it is actually a whole succession of storms clumped together.

When you see the anvil-shaped head of a mature storm cell, as in photo 10, that bit of the storm is past its prime. It is the apparently lesser storm clouds – those with knobbly tops that have not grown tall enough to have an anvil top (such as the one growing under the anvil on the right of photo 10) – that are producing the rain, hail, thunder and lightning to keep the storm going.

Is the storm moving around?

There's a widely held belief that thunderstorms move round, so ones that have passed come back somewhat later. It may seem like that when you breathe a sigh of relief as the current storm cell drifts away. Then, apprehensively, you hear the thunder approaching again.

It is a consequence of the cell theory that new cells can generate where the old ones originally appeared to come from. The old storm cell, to which you've just said goodbye, won't rejuvenate but there are other cells being born and growing to maturity almost anywhere when a major heat thunderstorm erupts. It will be one of these that now threatens. So the storm has not moved round you – you are just in a large area of storms, which means that a new cell or cells can come along to replace the ones that have dissipated.

When can we expect single storms?

There are many more single-cell storms dotted throughout the year than there are multi-cell summer ones. These individual storms come along in cool, very unstable airstreams and often in the spring (photo 23).

Photo 39 An imminent thunderstorm will often form a 'roll cloud' just ahead of it. When this roll passes, the storm arrives – expect very heavy rain and hail as well as thunder and lightning.

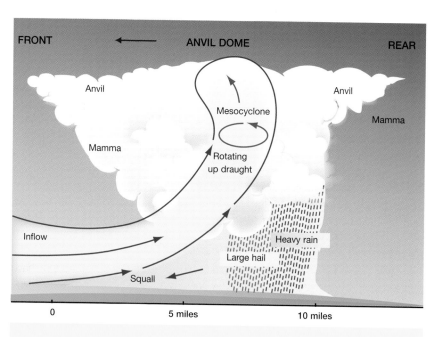

Fig 11.2 The anatomy of a supercell is different from a normal thunderstorm cell. The air that is drawn in towards the storm contributes to a massive up draught. This up draught develops rotation and becomes a mesocyclone, which can develop into a tornado. It is the complex nature of the up draught that leads to very big hailstones. As the supercell passes, so the hail becomes rain and then tails away. Mamma (see photo 61) is often seen on both the leading and trailing edges of the anvil cloud.

They are usually mixed in with big showers that are not themselves likely to be tall enough to produce thunder. The shower clouds that produce thunder may eventually acquire anvil-shaped tops but before they do so they will usually produce a few rumbles of thunder (which means a few flashes of lightning) and will pass. The big heat storm described above often stays around because there is little wind to move it. Single-cell storms will usually blow by on moderate – or even fresh – winds.

How do we get thunder in winter?

Here's a scenario. It is an afternoon of nasty winter weather. We are glad to be indoors as wind and occasional rain make outdoors very uninviting. However, it is not all that cold by winter standards. Then the sudden dramatic change described on page 84 occurs.

Such wild weather has to be due to the passage of a sharp cold front – and cold fronts are about the only weather system that will give you thunder in winter.

This applies over the land but not over the sea. Cold air from Arctic or polar regions, as it is brought south on north or northwest winds, finds itself over increasingly warm sea. Cold air over warmer sea is a recipe for the growth of great shower clouds that erupt in vast fleets over the ocean wastes. Coastal regions that face the wind will feel the brunt of these showers, which will be of rain, snow and hail and will sometimes grow big enough to produce thunder. This applies not only to winter but also to spring.

What's a supercell storm?

There is a great beast of a storm called a *supercell*. Luckily they are not very prevalent in Europe but when they occur they leave a lasting mark on those they affect. A supercell is a single cell but much vaster in extent than the thunderstorm cells described above. It is a great storm that comes up against the wind. However, that wind is of the storm's own making for it feeds the coming storm's updraughts. It has been recognised for a long time that it is the worst storms that come up against the wind (see Fig 11.2).

Previously there has been a pregnant hush in the hot, humid air with hardly a breath. Then a wind springs up directed towards the storm whose thunder you can hear in the distance. This wind does not usually get above moderate in strength but it is a good harbinger of a bad storm to come.

By the time the storm arches over you, the wind will have died away but is replaced by savage gusts from the direction of the storm. Lightning and thunder are now close.

If you suspect that the coming storm is a supercell then allow for hailstones at least as big as golf balls – but as big as tennis balls in the worst cases. These can be very dangerous to humans and animals. Cars will be dented and skylights shattered. Livestock in fields can be driven mad by the bombardment and should not be approached.

It may have happened years ago but you will remember when you've met a supercell. The one I experienced in northeast Essex in 1958 started in the Solent area and cut a swathe through Surrey and Berkshire before moving northeast to wash people out of their homes in Chelmsford. It became known as the 'Wokingham storm' as the hail was at its largest there but it managed to retain enough energy to blow in a window of the cottage where we were living at the time. It was one of the first to be recognised by meteorologists as one vast single cell rather than an agglomeration of smaller cells. In the Great Plains area of the United States, supercells are endemic and are the breeding grounds of tornadoes.

Are there special forms of cloud that presage thunder?

There are certain storms that are different to the ones described above. They are often very spectacular with lightning that is so vivid and continuous that you can see to read by it. At the same time the thunder is also more or less continuous. Yet they may not be as dangerous as they are formed thousands of feet above your head.

When storms of this high-level variety are destined for your area, there are special kinds of clouds that foretell their coming. Once you come to recognise these cloud formations you will have a valuable tool that will warn of possible storms later.

The two special cloud types that run ahead of high-level storms are formed in the medium layers and so are alto clouds. They also form in lines along the wind at their level and one (*altocumulus castellanus*) grows little mounds of cloud that look like battlements (hence the name) along its length (photo 41). The other (*altocumulus floccus*) looks like a flock of fat woolly sheep grazing in the blue field of the sky (photo below). So both names, castellanus and floccus, remind us of what they look like.

Photo 40 It is summer and thunder is forecast in the clouds. These fragmented lumps of medium-level cloud resemble the flock used to stuff mattresses (or a flock of sheep grazing in a blue field) and so are called altocumulus floccus. They form in lines along the wind at their level, ahead of thundery outbreaks – but it may not necessarily thunder where you are.

Strange hail

Not all hail is in the shape of little spheres. On the afternoon of 17 May 1993 almost circular plates of ice fell at Woodlands St Mary, Berkshire. They were about half an inch (12mm) across and 2mm thick. The plates were smooth and crossed by four spoke-like transparent regions.

However, they were not the first. On 15 May 1664 hailstones, 'flat and rough like fritters and broad as a half-a-crowne' fell in Oxford, while in September 1930 someone climbing Mount Olympus in Greece observed hailstones shaped 'rather like the underside of a button and the size of the circumference of a penny'. At Burton on Trent in July 1975, flat ovoids of ice fell during a storm.

Supercell storms have such strong updraughts that they can support very large hailstones which can melt and fuse together, leading to great knobbly stones several inches across. Such stones are well-known on the Great Plains of North America but I have personally handled similar stones in southern England which fell on the afternoon of Sunday 5 June 1983 in a great swathe stretching from Weymouth in Dorset to Chichester in Sussex. This region has a lot of market gardening and the damage to greenhouses was immense. These stones came from a series of supercells whose great gusts led to many yachts capsizing in the Solent.

The biggest stone measured scientifically fell at Koffeyville, Kansas on 3 September 1970. It weighed 1.67 pounds (0.76 kg) and was 5.6 inches (14cm) in diameter.

Hailstorms have changed history. In the Book of Joshua we read 'As they ran from Israel... The Eternal rained huge hailstones from heaven on them ... they died of these. Indeed more died by the hailstones than at the hands of Israel sword.'

In 1360 goose-egg-sized hailstones killed hundreds, if not thousands of King Edward III's men and horses near Paris. This so demoralised him that he agreed to sign the Treaty of Brétigny.

These clouds move in, sometimes quite early in the day and – if you are at risk from the coming storms – at right angles to the wind you feel. For instance, the wind may well be easterly but the lines of floccus and castellanus moving in are coming from the south. If, on the other hand, the clouds move in much the same direction as the wind, then – while the sky may cloud up – someone else is likely to get the storms. When this kind of high-level thunder is likely, the winds at the level of the clouds are going to be light. Thus we get oddities like long trails of falling cloud (called *virga*)

Photo 41 Another type of pre-thunder cloud, which also forms in lines, grows little turret-tops and so is called altocumulus castellanus. This cloud can come in on a wind whose direction is widely divergent from the wind you feel. When you see these cloud types, check the forecast for thunder later.

appearing to come almost straight down the sky. The visibility is often poor even up above you, where chaotic fragments of cloud often populate the sky.

What is a high-level storm?

This kind of storm is due to thunder clouds growing out of the layer clouds of a front and may occur something like two miles (3,000m) above your head. If you are flying at altitude you see them sprout like cauliflowers out of the white expanse of cloud that is the top of the front. They shoot bolts of lightning between themselves and sometimes these ribbons of lightning chase one another for miles across the sky. To the ground-based observer, the majority of the lightning is sheet lightning, which can be spectacular and creates excessive peals of thunder but will not hurt you. However, when there are lightning strikes to earth these can be of the worst kind. Many of the most spectacular photos you see of lightning are due to high-level storms.

12 lightning

No one, to my knowledge, has a comprehensive explanation for how thunderstorm clouds become electrified. There are many theories, some of which are more plausible than others, but even these do not appear to fit all cases – so we will just look at the results and not delve into how they come about.

Thunder cells develop volumes of negative electric charge near their bases and regions of positive charge further up in the clouds (Fig 12.1).

Fig 12.1 Due to processes going on inside the cloud, negative electric charge gathers around the base of a thunder cloud while positive charge congregates further up. The negative base induces positive charge in the earth below. Thus there is a strong electric tension between cloud and ground, which breaks down when a lightning flash jumps the gap. Lightning conductors can dissipate the resulting electric current harmlessly to earth. Broadleaf trees, however, offer considerable resistance to the passage of the current and are therefore very often riven by the strike.

The voltages (electric pressure) that are induced amount to millions of volts and the difference between two opposite sets of charges is sufficient to create giant electric sparks that we perceive as lightning.

Lightning can occur between clouds and the ground or between two oppositely charged regions in the clouds themselves, in which case there is no lightning flash to ground at all. It is cloud-to-cloud flashes that give rise to the term *sheet lightning*.

Lightning strikes

The way lightning behaves when it strikes, say, a house is often incomprehensible. If the electric current can find a path to earth then often not much damage is done. However, if the bolt is thwarted and cannot find earth it will jump to any nearby conductor which may be items such as your TV, your washing machine or your electric heaters.

Recently a house close to where I live was struck. Almost every electrical appliance in the house was burned out or rendered useless. However, they were not the only ones to suffer. We think that this was a multiple strike because the modem of my computer was damaged, as well as that of the SKY TV; also a subsidiary strike occurred to overhead telephone lines nearby.

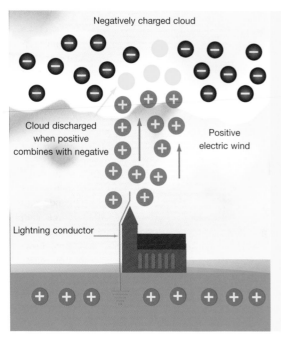

Negatively charged cloud

Cloud discharged when positive combines with negative

Positive electric wind

Lightning conductor

Fig 12.2 The action of a lightning conductor is two-fold. The point at the top of the conductor creates a positive 'electric wind' that streams up into the base of the cloud and helps discharge the build-up of negative charge there, so inhibiting a lightning strike. However, should a strike occur, the conductor will conduct it harmlessly to earth.

Staying safe

Take care to get out of the pool when lightning is imminent or even only possible. Water conducts the current of the flash but may use you as a path to the water. The results can be fatal. On Sunday 22 July 2006 a father on holiday in Tuscany with his wife and three children died instantly when violent storms broke out. He was sitting with his feet dangling in the water when the bolt struck. Danger did not seem imminent as the family had only heard distant thunder but even so they were about to take shelter as the blue sky had suddenly turned inky black. If swimming in the sea, while the risk is much less, get ashore and dry off. Seek shelter but not under cliff overhangs.

The telephone wire is a potential path for damaging voltages to enter your house and that is how the lightning attacked us. When storms threaten, it is important to disconnect the TV aerial and the modem and ensure that the two connections you now have are separated by a good gap so that it is unlikely that the lightning can jump across.

Where lightning strikes, it creates earth currents that can do damage to other installations and so you do not have to be struck directly in order to experience damage. Keeping clear of earth currents is important if caught in a storm in the mountains (see Chapter 15).

Photos of lightning show how the main branch of the flash is surrounded by other lesser forks and why a flash of lightning is a complex affair, often impossible to explain in detail.

When lightning occurs some way away, the accompanying thunder is of the kind that peals about the sky. However, if the strike is close, only a loud bang and sizzle is heard as the sound waves, generated by the heating of the channel along which the lightning has travelled, will all have gone elsewhere.

Is sheet lightning dangerous?

When a lightning flash occurs between clouds, the light generated is often reflected from clouds surrounding the flash. You sometimes don't even see the flash – only the reflection that spreads the light into a sheet. As this occurs often a mile or two above the ground and does not come to earth, it is said that sheet lightning is not dangerous. However, remember that these high-level storms can also shoot bolts to earth, but often, when you can see the sheet lightning, the storm is too far away for it to affect you directly – not that anyone can say categorically that it will not, as there may be other storm cells developing above your head.

What is ball lightning?

There is a particularly odd form of lightning where a ball of light forms – often not far from the ground – and obtains an independent existence, bowling along at about the height where it formed. It can move against the wind and can enter rooms by the expedient of forming a hole in a window pane, moving through and reforming within the room. Such balls may disappear with a bang when they make contact with something metallic in contact with earth, such as a kitchen range. There is no consistent theory of how ball lightning forms and how it maintains itself during its life, but if you should see it keep out of its way and join a very select band of observers who have actually witnessed this curious phenomenon. I personally have only known one person, a German called Horst Müller, who has seen ball lightning and he observed it at a location on the North German Plain – an area renowned for its hefty thunderstorms.

What is summer lightning?

This may not be a universal term but it is one I learned from my mother, who grew up in Hastings. On page 95 we talk about the high-level electric storms that sometimes drift up across the English Channel from France. These generate a large number of cloud-to-cloud lightning flashes and – being less than 50 miles (80km) away and very high up – they can be seen after dark over the coasts of France. They are too far away for their thunder to be heard and all you see on a summer's night as you stroll along the promenade is the lightning playing around the distant cloud tops. This is *summer lightning*.

We do not have to be on the south coast of England to see this kind of lightning but you are more likely to notice it in seaside places.

What is thunder?

If you have ever been unlucky enough to be close to where a lightning flash strikes, then you will know that there is a loud crack – often accompanied by a hiss. Every time a flash of lightning occurs, the path of the flash gets exceptionally hot. This explosively expands the air around the flash and the result is thunder.

Thunder rolls around the sky by coming to you from different clouds that reflect the expanding air around the flash. It is particularly noisy when there are cloud-to-cloud flashes so a very noisy storm is not necessarily a dangerous one.

Remember that you cannot have thunder without first having a flash of lightning somewhere. So every time you hear thunder, even if you do not see any flashes, you know full well that the clouds are electrified enough to create lightning.

Are they coming your way? If you can see lightning then there is a way of finding out.

How far away is the storm?

Light travels instantaneously (in this context), while sound travels only at about 750mph (about 333 m/s), so the thunder from a flash arrives some time after you see the flash. This way we can find the distance of the flash – and so of the storm.

Count the seconds between flash and thunder and divide by 5 = distance of the flash in miles.

Do the same but divide by 3 to find the distance in kilometres.

Wait a minute or two to see if the distance becomes less. If it does, it is likely the storm will affect you directly. It is not absolutely certain because storms die out in one part only to rejuvenate in another. Sometimes the storm may be approaching but for some reason never actually gets as far as you.

Am I likely to be struck?

The answer is that no one can be sure but there are ways of diminishing the risk.

Firstly, indoors is the safest place to be. Well, not quite – you cannot get much safer than in a car. A car is a metal box, well-insulated from the earth, and I have never heard of a car being struck by lightning. It is even less likely if the car is moving.

Among the most dangerous places to be are open spaces like golf courses, especially if you are actually handling a golf club. Also do not have an umbrella up in a thunderstorm. It has metal spokes and other fitments and while there are those who say it does not matter if you are close to metal, it is better to be safe than sorry.

Yachts at sea are very rarely struck, which might seem odd as they have a long pole – often metal – sticking up towards the base of any thunder cloud that happens to be overhead. Many yachts are fitted with lightning conductors but even those that don't have them do not seem to fare any worse than those that do.

I have a theory about this. Just as moving cars seem to be immune to lightning strikes, I think it may be that the top of a mast is in constant motion when sailing and so is always preventing the build-up of charge. Not so the yacht in the marina or on the mooring. If the mast is still then you could get struck.

There's a famous line from the song *Pennies from Heaven*:

So if you hear it thunder, don't run under a tree.

Even if it means getting wet, this bit of homely advice is well worth taking note of. But it can depend on the tree. It is a fact that broad leaf trees like oaks, elms etc regularly get struck, whereas conifers hardly ever do. The reason would appear to be that conifers are covered in pointed needles and points produce an electrical effect that helps to prevent lightning discharges.

So in a thunderstorm, get into the nearest coniferous forest if you can. There is one practical drawback to this advice. Broadleaf trees will keep you much drier than any conifers except for perhaps cedar – but the best advice is to get wet rather than struck.

How do lightning conductors work ?

The function of a lightning conductor is two-fold. Firstly, it tends to attract a local flash so that it can be conducted safely to earth. Secondly, and possibly more importantly, it can discharge an overhead thunder cloud so that it will not produce a flash in the first place (Fig 12.2).

Lightning conductors are pointed, and effective ones are attached to a thick copper strip that is well grounded in the earth. The negative charge in the base of the cloud induces positive charge in the earth below and so also in the point of the lightning conductor. The effect in the latter is to create an electric wind of positive air molecules (ions). These stream up to the cloud and neutralise some of the charge in it, thus making it less likely that a strike will occur.

Anything in contact with earth that is pointed will produce the same effect as the lightning conductor and so make a strike less likely. Thus masses of pointed conifer needles will produce a big electric wind.

Anyone who wants to construct a lightning conductor must ensure that the thick copper strip they use is as straight as possible, with no sharp right-angled bends etc, and the metal that points towards the sky must be well-pointed and firmly soldered or brazed to the strip.

Who's firing the guns?

Once upon a time it was regular practice to fire big guns over ranges close to military and naval establishments. Now, in these days of missiles, it is very rare to hear big guns. So why is it that we can quite often hear what appears to be big guns? The answer is that what you are hearing is the sound of thunderstorms maybe as much as a hundred or more miles (160km) away.

When sound travels long distances, it loses the higher frequencies and so becomes a very gruff noise. This is why storms a long way away sound like big guns. What happens is that the thunder spreads skywards and becomes reflected off temperature inversions higher up. These refract the sound wave back down towards earth and, as thunder travelling near the ground dies out over a distance of ten miles or so. This leaves the skyray waves to be heard even though people much nearer to the storm hear nothing.

13 fog, mist and wet air

If you ask anyone what makes air, they'll invariably say oxygen and nitrogen but forget to add the one that makes weather – water vapour. However, any parcel of air will contain a quantity of water molecules mixed in with the nitrogen and oxygen. Even air over deserts is not entirely dry.

The important thing about water vapour is that it does not take much to turn it to water and rather less to turn the water to ice. If there were no water vapour, there would be no clouds, no rain, no snow, no mist or fog and definitely no dew. In fact weather, as we know it, would not exist, just as it does not exist on the moon.

Does humidity increase before rain?

According to many species in the plant world it certainly does.

> Of pimpernel whose brilliant flower
> Closes against the approaching shower
> Warning the swain to sheltering bower
> From humid air secure.

The dandelion closes up before rain as do convolvulus, chickweed, clover and tulips, but the pitcher plant opens its mouth wider. Fungi sprout in abundance when conditions are damp and likely to remain so.

Where does water vapour come from?

The short answer is from the oceans, with a small input from lakes, rivers etc. Water surfaces are always losing water molecules and the bigger their area, and the warmer they are, the more they contribute to the wetness of the air flowing over them. Vast amounts of water vapour escape into the atmosphere from the oceans and seas and form the fuel that will energise clouds, rain, hail, snow etc.

When water evaporates, it takes heat energy from the water and this energy is stored in the vapour molecules. When the vapour turns back into water again, as it does when clouds form, this heat is released. This is the most powerful way for heat to be transported to higher altitudes and is a major source of inversions.

What is relative humidity?

Humidity is a very difficult thing to measure. It is not like temperature, where to measure the degree of heat in the air you can just use the amount by which a liquid such as alcohol or mercury expands up a very narrow tube. Humidity depends on temperature and pressure and we have to resort to measuring it against the amount of moisture in the air that could result in fog. Hence the 'relative' in relative humidity.

To explain relative humidity we have to confine a specimen of wet air in, say, a bottle. Keep the temperature constant. There will be a certain amount of water vapour in the air in the bottle. Now put in a drop or so of water. These water drops will evaporate in the bottle and increase the amount of vapour. When we see the vapour just begin to condense, ie little water droplets form on the inside surfaces of the bottle, the air has as much water vapour as it can hold at that temperature. We say the air is saturated with water vapour and we call this amount 100 per cent. But originally the air (before we added more vapour) had maybe only 60 per cent of the amount that would saturate it and so the relative humidity was 60 per cent.

Now, repeat the experiment from the beginning but do not put in any extra water. Just pop the bottle in the fridge. The inside of the bottle becomes misty with condensed water droplets. This is because the colder the air, the less water vapour it can hold and so some has to condense into water, which it does on the inside surface of the bottle.

Take the bottle out into the room and the mistiness disappears as the warmed-up air can absorb the extra water vapour. Its RH is now less than 100 per cent ie it is not saturated.

Why is it so sticky-hot?

When we get a hot spell the air may be relatively dry or it may be humid. In the latter case the RH may be 80 per cent or so. The human body will attempt to keep itself cool by sweating, which means it will release moisture through the sweat glands. This moisture will evaporate, especially when it is hot.

Evaporation needs heat energy to allow the water vapour to escape from the sweat, which will try to hold on to it. This heat energy is taken from the skin so that sweating cools the body. However, if the RH is high the net amount of water vapour that evaporates will be lower and so the skin will be less cooled and we feel hotter. In the extreme case of 100 per cent RH just as many vapour molecules return to the skin as escape from it and so there is no cooling effect. This is what we get in rainforests and other oppressively humid tropical places.

How does wind affect body temperature?

If we get a wind blowing over the body, the evaporating vapour is carried away allowing more to escape from the skin and we feel much cooler. The air need not be any cooler than it was before the wind blew but we will feel cooler. Cover the body in artificial sweat, as we do when we come out of the water after a swim, and despite the air temperature being quite high we can feel cool or even cold because of the heat being taken from our bodies as the water evaporates.

When the temperature is already low, evaporation of sweat makes us feel colder. Thus we experience wind chill so that even though the air temperature is above freezing it feels, in the wind, as if it is well below freezing. If you get saturated by a shower or by having been immersed in water, the effect may be so severe as to lead to exposure and the loss of deep-body heat. This can, in the worst cases, lead to death.

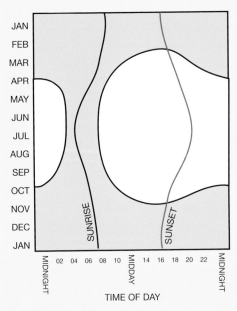

THICK FOG AT LIVERPOOL AIRPORT BY SEASON AND TIME OF DAY (AFTER G J BINDON)

Fig 13.1 How thick fog distributes itself around the day and season at a typical low-level place: Liverpool Airport. We see that fog occurs around dawn throughout the year but it is only over the mid-winter period that it will bedevil the home-going rush-hour. From late spring through the summer and into autumn it is very rare for there to be fog during the day, and in summer thick fog rarely forms before the early hours of the morning. Between November and March fog may occur in most places. However, there will be some places that, because they are low-lying or for some other reason, are particularly prone to fog. Do not lose sight of the fact that lesser fogs can occur at other times and they may cause temporary disruption.

The wind chill chart (Fig 4.1) allows us to assess how it will feel to the exposed body when the wind blows at a certain speed and the temperature is a certain value. For example the air temperature is 10°C (50°F) and the wind speed is 15 knots. It is going to feel cool. If the wind speed increases to 20 knots then we will feel cold. Taking another example, with the air temperature just above freezing (2°C/36°F), if we go out in a strong wind (force 6/7) it will feel icy cold. So long as the winds are light (force 1–3) it will just feel extremely cold when the temperature is -10°C (14°F) but should the wind get up to 10–15 knots it becomes dangerously cold and with a wind of 20 knots or more there is risk of frostbite.

What about animals?

Only hairless humans can profit from the cooling effects of sweating. Furry or hairy animals may sweat – horses do particularly – but their coats create a micro-climate just over their skin where it is windless and the cooling effects of evaporation are minimised. Dogs are particularly badly served as they can only lose moisture from their tongues, which is why dogs pant in hot weather. It is really only humans to which wind chill effects apply.

Why is there dew?

In olden times people did not understand why, overnight, the grass, plants etc became wet. The very word dew seems to involve the deity – dew came from God. After all, it hadn't rained so where had the water come from?

We now know that before water drops can form there has to be something solid for them to form round. In Chapter 5 we identified these condensation nuclei as dust, chemicals, sea salt from waves etc, but what are the condensation nuclei for dew?

Put a blade of grass under a microscope and you will see that it is covered in what look like hairs (cilia). Many other plants will be the same and it is the tips of these hair-like features that provide the minute solid centres on which water molecules can condense and so form dewdrops.

Photo 42 Dewdrops gather on the little hairs on the surfaces of vegetation. The night needs to be cool, but not frosty.

We've already seen that when you lower the temperature of a parcel of wet air, it has to shed vapour by forming water drops. This is what happens when the ground cools in the evening and overnight. The air condenses its excess water vapour as dew.

WHAT DOES DEW TELL US?

The formation of dew is a good indicator of how humid or otherwise the air is. It can help to back up the reading of a device for measuring humidity – a hygrometer – which you can buy from a garden centre. These are usually dial instruments and they are not very accurate – but then the majority of people do not need high accuracy. They just want to know if the air is more than normally humid – or maybe more than normally dry. If you want to measure relative humidity accurately, you have to invest in a device called a whirling psychrometer and very few people will want, or need, to do that.

Nights of dew go with starry or moonlit skies because these provide the conditions for radiation and thus the rapid cooling of objects in close contact with the ground. Such nights often indicate fair days to follow.

> *When the dew is on the grass*
> *Rain will never come to pass.*

On the other hand, a morning without dew often means that it has been cloudy overnight, thus preventing the ground cooling. Such cloudiness may well run ahead of a warm front.

> *When grass is dry at morning light*
> *Look for rain before the night.*

Photo 43 Another favourite on which dewdrops may condense is a spider's web. The dew throws into relief its astonishing intricacy.

Fog and the seasons

The two things that contribute to fog are cooling of the air and increase in the RH. This is why fog occurs mainly in the evening and morning as well as overnight. Incidence of fog also depends on the season.

The earth beneath our feet is a storage heater – not as big a storage heater as the oceans, but it is one nevertheless. During the late spring and summer the earth absorbs the heat of the sun; this heat penetrates deeper and deeper until late summer. Then the cooler air of autumn arrives and now the heat stored during the summer begins to give itself back, so that on many occasions the air becomes warmer than you might expect. The result, when winds are relatively light, is for warm air a few thousand feet aloft to exist over cooler air near the surface. This produces what is called a *stable* situation and it is the same situation near the ground that is called an inversion (page 13). In inversions, the tendency is for the warm air layer to try to sink towards the cooler one. Autumn readily creates these conditions and fogs and mists as illustrated below.

This is in direct contrast to spring when, after the winter, the ground has lost its deep heat and little heat is being conducted from below to the surface. However, the sun is daily gaining altitude and so starts to heat the ground by day. The seas are relatively cold so very often cool or cold air arrives over heated ground. Now we have cool air over warm ground – the exact opposite of autumn. This is an unstable situation and is a big contributor to the ebullience of spring as the alabaster shower clouds climb high in the sky and showers are very prevalent. There will be a lower chance of fog in spring over land although this may not apply to the sea.

Photo 44 The rising sun throws rays through the trees and illuminates the morning ground mist. This kind of mist can occur at most times of year but is most prevalent in autumn and spring. It may cause some problems on country roads in the early morning.

Fog by day

The power of the sun and the warmth of the ground during the summer mean that thick fog is rare except in the period surrounding dawn. Early-morning fogs disappear very rapidly in spring and summer, as Fig 13.1 proves.

When the evening sky clears and dew indicates that the air is already humid, then fog is very likely. A light wind also contributes to the incidence of fog.

As the ground cools through losing heat by radiation to space, it is easy for the RH near the ground to grow to greater than 100 per cent. This now means that the air has to shed water vapour. Its only means of doing this is either by condensing into dew on the grass etc or by forming water droplets, ie creating mists or fog. These fogs are called *radiation fogs* and Fig 13.1 shows when they are most likely.

However, there is another way in which it can become foggy – and rather suddenly at that. This is when a fog bank is carried over you by the wind. This is called *advection fog*. As convection is transporting air in the vertical, so advection is transporting air in the horizontal (photo 44).

If you live near the coast or go to sea then you will experience sea fog. Sea fogs depend on the relative temperature of a mass of air flowing over the sea to that of the sea below it.

What about sea fog?

Fog over the sea is the result of warm, humid air arriving over colder water. It does not vary with time of day but it does with the season. Unfortunately for those who sail or cruise coastal waters, spring and early summer are the times when sea fog is most likely to occur. These are also the times when seabreezes are at their strongest, and sometimes a warm morning on the beach under a sun that is very high in the sky is suddenly ruined by the arrival from seaward of damp, clammy fog. This fog is brought in on newly created seabreezes but it is a phenomenon of coastal regions and the fog does not usually survive more than a few miles from the main sea coast.

It is a menace because it may well not be forecast and inshore waters are notorious for hazards. When a small craft suddenly loses sight of the land it may become endangered because it cannot get its bearings. Sailboarders who have strayed off-shore are even worse off because they they may lose their bearings towards the beach and may instead head out to sea.

Southwest winds are the most likely candidates for sea fog but sometimes this fog does not show itself until the air lifts over the actual coast itself. Coastal hills will then become shrouded in fog that can persist all day until either the airstream gets drier or the wind shifts.

Photo 45 A wind lifts fog up the little valley. If the hills are not too high, their tops may well be in sunshine while the valleys below are lost in fog.

Clear airstreams that have to traverse a long passage over cool water from a land mass, after 100 miles (150km) of sea surface to cross, end up fog-bound on the opposite shore. A good example is northeast winds in the North Sea. The wind leaves the coast of Denmark relatively dry but acquires more and more moisture for every mile it travels. By the time it reaches the east coast of England it can be dense with fog.

However, if the wind shifts to the east, the much shorter passage to England does not allow the air enough time to become saturated with water vapour, so there is no fog risk. This may not apply to coasts further north because here an easterly wind has to travel much further between coasts than it does further south.

Anticipating sea fog

To assess the likelihood of sea fog, consult coastal waters forecasts or shipping forecasts. If a ship's funnel or other smoke hugs the sea surface then fog is likely. If you see some cumulus clouds about over the sea, you can forget about fog. However, if the cumulus is blowing off the land then sea fog is possible. It may be unlikely but you cannot be sure.

Oddities of fog

When you are in a clamp of fog you may not be able to believe your senses. Fog absorbs sound, or diverts it – reflects it, if you like. The sound of a fog signal or some other noise may come to you from some direction that is not necessarily where the source of the sound lies. Low-frequency sounds travel further in fog than high-frequency ones, which is why some fog signals grunt. Also, because the air becomes stratified in foggy conditions, fog signals or other sounds become bent (refracted) and may be lost, disappearing upwards before they ever reach you.

Things suddenly appear out of a fog and if you are driving a car or a small boat into a continuously white sheet of fog you can get what is called *white field myopia*. This means that you become deprived of your ability to judge your position relative to objects that may be in your path. You become mesmerised by constantly peering into the featureless wall of fog which makes the chances of hitting something that much greater.

What is mist?

Mist is, in reality, just thin fog, but you can get mist when you would not get fog. Mist can form at sundown because of the speed with which the evening is cooling down, but later the lights twinkle brightly, showing that the mistiness was only temporary.

Those who have to issue forecasts for the general public in Britain (and it may well be different in other countries) have decided that if you can see between 200 and 1,100 yards (183 and 1,005m) then you have mist and if it is under 200 yards (183m) you have fog. However, dense fog is less than 50 yards (46m) and these limits are laid down mainly for car drivers.

At sea, fog is present if the visibility is below 1km (2,200 yards); it is described as poor between a kilometre and 2 nautical miles.

What is smog?

Smog is short for smoke fog and it is luckily less prevalent today than it was 50 or more years ago. However, it still occurs when winds are light and fog forms in association with effluent from industrial processes, to which we have to add diesel fumes and car exhausts.

Four thousand people died in the great London smog of 5–9 December 1952 and this tragedy led to the Clean Air Acts, which effectively prevented people in towns from burning fossil fuels.

For a smog you need to be near the centre of a strong anticyclone so that there is next to no wind. A strong anticyclone usually creates a thermal lid (an inversion) maybe 2,000 feet (610m) or so from the ground. Sometimes, however, this inversion is half as low again. This prevents convection currents and so any smoke, fumes etc get trapped below the inversion and cannot escape. They build up with time and, unless wind appears to drive the smog away, develop to densities that are dangerous to people with respiratory complaints and even to otherwise healthy people.

Smogs occur habitually in some places. Los Angeles is renowned for its smog.

What clears fog?

Most morning fogs clear from the bottom upwards. If fog is fairly bright it means there is no cloud layer above and so the sun can produce enough radiation to clear the fog. What happens is that, even though you cannot actually see the sun, its heat radiation penetrates the fog layer, warms the ground and sets off convection currents that mix the wet air near the ground with drier air further up. This also increases the wind speed (Chapter 2).

If a warm front or occlusion is approaching, it often brings drier air ahead of it and this can clear a fog. For a time, as the fog clears, there will be fumey lumps of heap cloud about until they, in their turn, disappear leaving sunshine. The fog may be slow to clear from some hollows.

If the fog is dark, it indicates either a very deep layer of fog or that there is a cloud layer above the fog. In these cases, the fog may take all the morning to clear or may nor clear at all.

Thick fogs will often be cleared by a wind getting up or, like hill fogs, will only clear when the wet airstream changes. Either the wind will shift direction to a drier quarter or the air itself will become drier with time. Sometimes both things happen.

Valley fogs often leave the hilltops in sunshine. This is when, if you should be up there, you will be rewarded with the unforgettable sight of church steeples, power lines etc sticking up incongruously out of the brilliant white fog layer that is filling the valleys. Winds that blow uphill (*anabatic winds*) occur in these conditions and contribute to the clearance of the valley fog.

Sea fogs can persist for days when the airstream continues from a warm, humid direction but, just as above, they will clear when the wind shifts. Sea fogs that drift inland overnight can clear over the land but the beaches etc can still be miserable.

14 sea weather

Unlike over the land, there is no daily change of temperature over the sea. The deep-sea temperature only changes slowly.

Depending on the season, the edge water can get much warmer or colder than the offshore sea. Along the shores, when the tide floods in over warm sand and shingle during summer days, the evening can be one of the best times of day to swim. On the other hand, the tide will come in well chilled in the early morning after a clear winter's night when the exposed mudflats, sand bars etc have become very cold.

So creek and estuary waters can vary considerably in temperature depending on the season and on what time of day the flood occurs. In sunny spring weather swimming off the beach becomes a possibility because the edge water, picking up heat from the sand and shingle, can get so warm.

Offshore, deep-water temperatures are at their lowest around the end of February/beginning of March and they climb steadily until late August/beginning of September, after which they decrease but still manage to give the countries of Atlantic Europe a warm autumn. The fact that January and February are the coldest months of the year is not only due to the sun being low in the sky but also to the fact that winds that blow from the seas are, possibly, losing heat to the cold-water surfaces.

Why does the sea sometimes freeze?

Occasionally in temperate latitudes the weather gets cold enough for the sea to freeze – not the deep sea, but the creek and estuary edges. You have to go to Scandinavia before you can expect the sea to freeze on a regular basis.

Pure water freezes at 0°C (32°F) but you need to cool dirty water to below zero before it will freeze. Seawater is full of salt and mud particles etc so we need a very cold snap for it to freeze; then it will be clear, freezing nights, when the tide floods late in the night, that will lower the edge water several degrees below zero. Then it turns to ice.

What about clouds over the sea?

In winter, when there is clear weather, there can be no cloud at all over the land but much cloud over the sea. This may be benign cumulus cloud or it can have grown into cumulonimbus and so be producing wintry showers of rain, snow and sleet.

You can get this sea weather invading the land when the wind blows on-shore (note: 'on-shore' means from sea to land while 'onshore' means over the land).

Let the wind turn round and blow from off-shore (ie from land to sea) and the coastal waters will be as free of cloud as the coastal lands are. However, as the air goes further offshore it will acquire moisture and so erupt into heap clouds.

Why do coasts get so much snow?

Cold winds from the Arctic regions sometimes come south when high pressure exists out to the west. If these winds travel over land, they may not have enough moisture to produce snow showers. However, if they leave a land mass and blow for a while over the sea, snow showers are the result. The sea will always be warmer than Arctic air and so it will soon breed showers.

This is why areas of land that jut out into the sea are very prone to snow showers, whereas other places not far away are relatively free of snow.

Sometimes the cold airstream that blows on-shore will transport its snow showers to the coastal lands. However, once the cloud loses its source of moisture, ie the sea, the snow showers cease. So the coasts get snow, whereas further inland there is little or none.

Photo 46 Occasionally, in periods of extremely cold weather, creeks and harbours freeze over. In the area of the English Channel and southern North Sea, this is very rare. This is Essex in the extreme winter of 1963.

Why is it often much windier over the sea than over the land?

The first reason is that even a rough sea is much smoother than the surface of the land. So the surface wind blows stronger over the sea.

Another reason is the time of day and the way the wind blows compared to the coast. Let's take the scenario when a wind blows more or less parallel to a coastline and it is evening.

We already know the wind mutes with evening over the land but what we must remember is that the wind over the adjacent sea speeds up to compensate. So we find that when the wind is east and blows along a south-facing coast, the wind just off the coast may double in speed compared to its speed over the land.

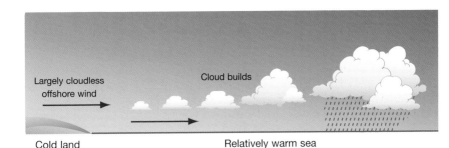

Fig 14.1a Mainly during winter, cold air blows off the land onto water that is warmer. The further this air travels seaward, the more chance that showers will erupt.

Fig 14.1b When the reverse situation occurs, and relatively warm air blows in off the sea over cold land, the cloudiness over the sea will disperse as you travel inland.

This doubling is the greatest increase you are likely to get but nevertheless the wind over the sea at night is almost always going to be stronger than over the land. Sailors who decide to go to sea overnight will often be disagreeably surprised at how strong the wind is offshore.

Seabreezes

When the isobars of a high-pressure region spread across a coastal sea area then the winds in the mornings are light and there is usually sunshine. The result is extensive seabreezes, which on the best days take over the wind patterns out to tens of miles offshore and similar, or greater distances inland.

In the latitudes of Atlantic Europe the seabreezes are strongest in late May and early June. This is when there is the greatest contrast between the sea and air temperatures and when, for instance, breezes that start from the south coast of England in the morning are still rolling slowly inland when they reach the Thames Valley in the evening. This scenario will be repeated inland from other coasts but the effect is most noticeable when the coast faces south or east.

On the quietest days the seabreezes will wholly take over the winds of quite extensive areas of sea such as the English Channel and the southern North Sea. Seabreezes blow onto opposing coasts and this makes for frustrating areas of calm in the centres of these waters.

The seabreeze season is from April to September in these climes.

Thunderstorms over the sea

Heat thunderstorms of the kind you get over hot areas of the land do not naturally occur over the sea. However, over the deep sea, miles from land, single-cell storms occur quite frequently as great cumulonimbus clouds grow in cool, moist airstreams as they travel over warming seas. Clumps of storms occur along fronts and may go on for many hours. However, there are many more lightning flashes from storms over the land than there are from those over the sea – ten times as many in fact. This is probably due to the clouds not growing so high over the sea as they do over the land.

However, get into shallower waters and thunderstorms generated by heat over the land will invade sea areas and may be violent, especially when an old front intervenes.

As a case in point we looked at high-level storm outbreaks on page 95 and these will be just as virile over the sea as they were over the land. In these cases, nasty thunderstorms can invade leeward coasts and coastal higher ground may increase their severity.

15 hill and mountain weather

If you go up onto higher ground then you have to be careful. Weather goes bad in the hills and mountains much more rapidly than on the lowlands. Mornings that look fair can soon turn to low cloud, fog, drizzle and rain.

Why is weather worse over high ground?

For cloud and rain to occur, wet air needs to be lifted. Over the lowlands, air is lifted by convection but as soon as it goes high a new factor comes into play. Wind blowing up hill slopes is mechanically lifted so what may seem a fair day at low level, rapidly becomes bad higher up. Even if there is no precipitation, mountainous districts are much more likely to be cloudy than the lowlands (photo below). Then there is the fact that it gets colder. Temperature drops on average by half to one degree Celsius for every 300ft (100m) of ascent. So, if where you intend to go is, say, 1,000ft (300m) above sea level, the temperature can be some 3°C (5–6°F) lower than it was at low level.

Photo 47 When frontal weather sets in over mountains, the rising and sinking currents induced by the wind blowing over the ridges creates an unforgettable sky broken into lens-shaped elements, with much other cloud as well. If the lenses disappear and the whole mass becomes more solid, it may rain.

To this we have to add the increase in wind speed that occurs as you ascend. Together, decrease in temperature and increase in wind speed may bring you into the risky area of the wind chill chart (Fig 4.1). It seemed so pleasant lower down. Now things do not look so rosy. And there is another factor.

Air that cannot produce showers over low ground can get pushed into erupting into cumulonimbus over high ground. These clouds may get big enough to produce thunder with downpours and hail – catching out hill walkers.

How do I foretell possible deterioration?

The onset of high cirrus cloud, haloes about the sun etc as shown in photo 15, page 41, apply to the mountains because these clouds are too high to be greatly affected by even high mountains. It is clouds lower in the atmosphere than cirrus that will bring deterioration to the heights when the lowlands are still quite benign. If there are mountain summits in view, note whether any of them are producing snow plumes as the winds increase. Also note any dark clouds that are to windward and act accordingly. You cannot rely on the wind direction to tell you anything positive because winds try to flow through valleys and round hills and mountains so their direction may not be anything like the true wind direction. However, if you can get a reliable observation that the wind direction is backing, and the other sky signs are there, read this as prompting a return to the lowlands.

How do I allow for mountain weather?

Firstly, ask the locals what they think before you start. They know their area and will be happy to show off their knowledge. Always carry a knapsack with a thick jumper and a cagoule as well as waterproof trousers. Take some food for emergencies. Make sure you have a map and a compass.

Photo 48 In the mountains, once you have seen this remarkable formation, you won't forget it. This 'pile of plates' is a particular form of lenticular cloud and will normally remain stationary in the sky for a while at least. Photo: A. Gilkes.

What should I do if the weather suddenly closes in?

Firstly, do not do anything until you've thought about it. If the path is well-worn it's maybe best to follow it back down, where you may break into clearer air.

Certainly do not go on unless you are with someone who knows the area and thinks there is less risk in continuing than in turning back. If conditions become too bad, stay roughly where you are but try to keep moving about.

What should I do during thunderstorms?

Unfortunately, sheltering close to rock formations, especially cliffs and overhangs is the worst thing to do. Being closer to the base of a thunder cloud means that lightning is more frightening and the first instinct is to seek shelter.

However, sheltering close to cairns, rocky outcrops etc is unwise as these are favourite points for lightning to strike. When you have a lightning strike onto rocks etc above you, the massive electric current that is generated has to go somewhere. It runs down the vertical and near-vertical faces to find earth. If you are sheltering in the sharp bend where rocks and ground meet, then you can become a useful path to earth for the current. Thus, despite the deluge and the cold your best bet is to sit at least 150ft (45m) from the rocks. Sit hunched with your knees drawn up and your head on your knees. This way, you may be wet and cold but you are relatively safe.

Also remember that storms produce deluges. These do not have to occur where you are for them to be dangerous. Look up to higher ground. Could a flash flood rip

Photo 49 The steep undulations in mountainous districts can easily induce otherwise innocuous clouds to produce showers as is happening here. The bright horizon shows that the rain will not be long-lived.

down this hillside, releasing boulders or causing mud-slides? Or could this gully, which seems to provide such good shelter, become a raging torrent? This might take an hour or more to arrive after the storm that started it has moved on.

Can I tell if I'm in danger of being struck?

Although I've never experienced it, it is said that you can feel the build-up of voltage when your skin feels as if it is being touched by a spider's web. Another sign is when your scalp tingles. Way posts or other upright wooden stakes will sometimes exhibit St Elmo's fire, a blue coronal discharge that occurs around the top of posts etc. It was well known in the days of sailing ships when, in thunderstorms, St Elmo's fire played around the tall masts.

If you have a tent with metal poles, collapse it or keep away from it. Lightning shows a preference for metal things that are upright and well connected to earth. If metal posts are upright then they sometimes 'sing' before lightning strikes.

Special clouds in hills and mountains

There is a form of cloud that imitates the shape of the rounded backs of hills and small mountains. This cloud is called *altocumulus lenticularis* because it forms itself into lens-shaped elements that lie across the wind direction at their level. It is formed by big wave motions in the air. Where the air rises into the top of a wave it forms cloud, and when it sinks again the cloud disappears (Fig 15.1).

One unusual aspect of lenticularis is that it often stays stationary in the sky. Other clouds move with the wind but lenticularis, being formed by standing waves in the atmosphere, often stays put and sometimes great sausage-shaped rolls develop across the wind with clear spaces in between.

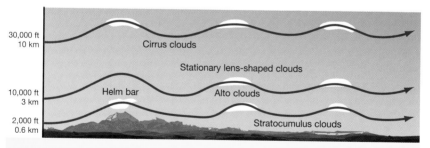

Fig 15.1 How stationary lens-shaped clouds form in stable airstreams downwind of a hill or mountain ridge. As cloud forms in rising air and disappears in sinking air, so the clouds mirror the waves in the air and may remain stationary for hours. These clouds can form right up to cirrus levels, while over the ridge itself a characteristic cloud called a *helm bar* is sometimes seen.

A pile of plates

An odd form of lenticular clouds is called a 'pile of plates'. Once you have seen this cloud form you will immediately know where the name has sprung from. At their best they do indeed resemble dinner plates stacked on top of one another (photo 48, page 117).

Interestingly, great shows of lenticular clouds may not lead immediately to rain although they are associated with deteriorating weather. By the time the low cloud and rain arrive, the lenticular clouds have disappeared (photo 47, page 116).

What is a föhn wind?

On the Alpine Foreland there occurs a special wind called a *föhn*. It is warm and dry and the wind is often strong and may reach gale force on occasions. The skies are often clear under föhn conditions.

For föhn to occur, large-scale winds must blow from the Mediterranean side of the Alps but föhn is just the local name for one of many such winds that blow over big mountain ranges. The similar wind that comes down the eastern faces of the Rockies is called the *chinook*. This means 'snow-eater' in Indian dialect because it can melt snow at a surprising rate. In Southern California it is the *Santa Ana* and in the Andes it is the *zonda* or *puelche*. In Serbia it is the *koshava* or *ijuka* and the *afganet* or *ibe* in central Asia.

Whatever the name, the föhn winds are always drier and warmer than the air on the windward side of the mountains. There does not have to be a local name because wherever winds from the western quadrants are forced to cross a mountain barrier, the area in direct lee will experience a more than usually balmy climate. Thus Aberdeen, for example, where the wind has to traverse the highlands of Scotland, has a form of these lee-side winds. Reverse the wind direction to north-east or east and places on the west coast of Scotland and northern England will enjoy the warmth.

One aspect of föhn should be mentioned and this is its known psychological effects. Hours or even days before a föhn wind blows, people who do excessively demanding jobs, such as surgeons, rearrange their timetables as it has been proved that they are more accident-prone at these times. It has also been proved that road accidents increase because of the effects that föhn winds have.

What about avalanches?

Despite all the research that has been carried out into avalanches, they are still the most feared of mountain hazards.

There are different kinds of avalanche. A 'loose-snow' avalanche is one that breaks away from the snow surface at a point where the snow has become unstable. It usually occurs only with snow that has fallen in the last 24 hours. It then descends along a pear-shaped track. A 'slab-sided' avalanche involves a whole area of snow moving, leaving behind a break-off wall. Sometimes only the upper layers move and the snow may be dry or wet. Avalanches may be airborne powder or they may flow, with the snow keeping contact with the ground.

Not enough can be said here to give you a comprehensive knowledge of the risk of avalanches, so listen to the snow forecasts, consult the locals and never venture out if there is a risk present. Do not, for instance, undertake any mountain expedition if a sudden change in the weather is forecast. Never start if föhn is blowing or is forecast and never start after a fresh fall of snow.

Steep lee slopes are the most dangerous, especially in changeable weather or after the sun has been out on the slopes for some time. South-facing slopes can be safe in the early morning but become dangerous when there is hot sunshine. Sometimes even the experienced guide may be unsure of the safety or otherwise of the conditions. Look for snow plumes being blown off mountaintops, as they indicate that the lee slopes will probably be deep in soft powder snow even though no new snow has fallen recently.

Photo 50 The wind is blowing from the west over the Cambrian mountains in Wales. Great standing waves are induced in the air. Where the air is rising, cloud forms and where it is sinking, the cloud disappears. The result is long, often lens-shaped ribbons of cloud stretched across the wind with clear slots between. As these clouds are built by the standing waves, they can remain stationary for some time.

16 freezing weather

When the days lengthen, The cold strengthens.

This really tells us what we already know – that January and February are the coldest months. They are also the months when the most severe storms are likely to occur.

In late autumn – up to the winter solstice (22 December) – the weather is riding on the backlog of the heat absorbed by the sea and land during the summer. After that – as the days begin to show signs of lengthening – the land is at its coldest and the sea nearly so.

Thus clear nights are almost bound to be frosty and precipitation is much more likely to be snow or sleet.

Why does it snow?

The fact that most rain starts as snow in the higher reaches of the clouds means that in winter, when it is freezing down to the ground, the snow never melts into raindrops as it does when it's warmer.

To understand about clouds and snow we need to appreciate that while water freezes at 0°C (32°F), the water droplets that form clouds can still stay liquid even when the temperature is well below the normal freezing point of water. In between the cloud droplets there is an 'atmosphere' of water vapour and also ice crystals.

The coldness of the upper air is relative. Sometimes the temperature at altitude is less cold than at other times. If, despite being well below freezing, the air at height is relatively warm, big snowflakes can grow because the air is also quite wet with water vapour.

The embryonic snowflake is an ice crystal onto which water vapour molecules attach themselves in a six-sided pattern. In an atmosphere relatively dense with vapour the snowflakes grow rapidly and can become quite large. These flakes now parachute down to a level where they can melt into raindrops. That is how rain forms.

Photo 51 It has been a very cold, clear night and there has been a sharp frost. The sunlight transforms the fern-like 'jack frost' patterns on the window into one of the most beautiful sights in nature.

Now, consider it is winter and the temperature throughout a snowflake's life is below freezing. Big snowflakes now cannot melt and so arrive on the ground – and maybe a lot of them.

Why does rising temperature presage snow?

In a run of cold weather we will get ridges of high pressure preceding the fronts of depressions. High pressure often means clear skies (or at least only partly cloudy ones) and clear skies mean very cold nights. When a front is in the offing – even before it has actually made its presence felt – the high-pressure regime is being replaced by a low-pressure one and invariably the weather warms up somewhat. As the temperature rises towards freezing, the risk of snow increases and the warming trend also means that the snow can be heavy and prolonged. This is well-known to Alpine and other peoples who inhabit mountains.

When the icy wind warms – expect snow storms.

Is it too cold for snow?

If the surface temperature is very cold, it is likely that the temperature up through the clouds is also colder than normal. Now there will be less chance of forming big snowflakes. Small, harder flakes are formed and sometimes you are surprised to hear the rattle of pellets of soft hail called *graupel*.

When it is grey and very cold you sometimes hear the remark, 'It is too cold for snow'. Now, it is actually never too cold for snow. What prompts this remark is probably the fact that very cold temperatures go with anticyclonic conditions when air is tending to sink rather than rise. The chances of snow are therefore very low.

Why won't the snow showers stop?

Much snow comes from showers rather than falling continuously but snow showers are not like rain showers. Rain showers stop and start pretty abruptly because the drops fall rapidly but snow takes time to fall and can often appear to fall out of a clear sky. This is because it is carried forward on winds that are moving faster than the shower clouds. This also explains why often the snow does not seem to stop between showers – it just gets less intense. The snow in the showers is often heavy and if it continued to fall like that it would soon lead to a foot of snow – but it does not. Every so often you get a glimpse of blue sky and you know that this is indeed only showers and not an Arctic blizzard.

What melts the snow?

Snow tends to melt from the ground upwards because the snow cover is colder than the deep ground temperature – and heat flows from where it is warmer to where it is colder. However, this is a slow process and the conditions that melt snow very rapidly are the arrival of air several degrees above freezing. Snow cover that lasts for days usually follows the passage of a cold front. The ridge of cold air that often follows a cold front leads to very cold starry nights and even the sun by day cannot do much to melt the snow.

However, a warm front eventually comes along. There may be more snow temporarily but the arrival of warm air means the snow melts rapidly from the top. Helped on by the melting at the bottom of the snow deck, the snow may thaw very rapidly.

This is not a pleasant period at all. There will be a raw feel to the day and pools and puddles form everywhere. Slush finds any imperfections in your footwear and fog and mist make driving difficult. It is such a contrast to the beauty and serenity of the recent snowy days that all we wish for is the dratted stuff to disappear as rapidly as possible.

Why doesn't all the snow melt?

Snow that packs into hollows develops its own micro-climate. Despite the general air temperature being several degrees above zero, the snow in hollows tends to maintain a sub-zero stratum of air over it so it refuses to melt. In mountain districts snowfields may survive right through the summer.

However, have you ever noticed that, when the snow is melting, parts of the pavements or roads keep their snow-cover longer than their surroundings? The reason for this is that the parts in question have recently been dug up. The in-filled trenches have enclosed air cavities, which insulate the snow from the deeper heat of the earth below and so it does not melt as readily as its surroundings.

Why have they got it wrong again?

Forecasters hate snow. If they have failed to forecast a smattering of light rain then no one complains, but get a few flakes of snow that have not been forecast and there is a chorus of complaint.

Sometimes there is no doubt about the fact that it will snow and forecasts reflect this certainty. However, the borderline between falling snow that melts to rain and falling snow that survives as snow is very blurred. If the temperature in the air decks near the surface just a degree or less higher, what they thought would be snow falls as rain. More alarmingly from the general public's point of view is the opposite when the air decks are colder than expected and un-forecast snow, which should have been rain, appears.

Photo 52 After a night of freezing fog, every branch and twig is coated with a covering of grainy hoar frost, producing one of the most enchanting sights of winter.

A further complication is how variable snowfall is. I remember one memorable occasion when we, living in a rural location, had almost 18in (45cm) of snow and it was impossible to get to work. In the local town where I worked they had an easily manageable 4in (10cm) and were reluctant to believe that I was not just taking the day off. Such isolated heavy snowfalls are rare but are often due to small local depressions – lows so small in extent and depth that they might not even appear on the weather charts.

What is a glazed frost?

If you experience a glazed frost you never forget it. I have only known two or three in my life and they are one of nature's worst nightmares.

Rare conditions of temperature in the air decks near the ground mean that the surface is below freezing but higher up falling snow just manages to turn to rain. However, the rain is only just above freezing itself and when it falls on any surface objects it immediately freezes, covering everything with a layer of ice. Leaves, boughs, gates – all become encased in a covering of ice that makes them look as if they have been cast in glass. More to the point, pavements are also covered in a sheet of clear ice, making normal walking nigh impossible, while roads become as slippery as skating rinks.

What is hoar frost?

Hoar frost is the white, grainy frost you find covering the ground on cold mornings. It is the icy equivalent of dew and every hair on plants close to the ground becomes a centre for the formation of ice crystals instead of water drops.

Photo 53 Fog that is freezing, or very close to it, will build up as rime on surfaces below freezing. Here, a wire-mesh fence makes a perfect base on which rime can form.

Sometimes on cold mornings you get fog and water droplets of the super-cooled fog immediately freeze when they contact solid objects. Thus the hoar covers the trees, creating one of the most beautiful sights of winter. It is this frost that builds up on the windward side of openwork fences etc when the freezing night wind blows. It is then called *rime*.

Why do icicles form on the south side of buildings?

Icicles form from the overhanging eaves of buildings when the combined effect of the sun and the heat rising from the building can just melt the snow on the roofs. The water so-formed is close to freezing and will perhaps only be liquid during the middle of the day. As the temperature falls with the afternoon the water, usually dripping from some projection or other, begins to freeze. The embryonic icicle freezes the water slipping down its sides and the icicle grows both in length and in width.

It is the sun, even in winter, that has the greatest effect in melting the snow on the roof, and so, because the north-facing side of the building does not see the sun, icicles rarely form there.

Photo 54 It is September and the air temperature is such that a ground frost cannot have occured, yet we seem to have a covering of hoar frost. The answer lies in millions of spiders' webs, upon which droplets of dew have condensed in the cool of early morning.

17 colours in the sky

Daffodils are yellow, roses are red, leaves are green and all because in sunlight they reflect their respective colours and absorb others. Yet look at them under yellow sodium street lights and only the daffodils will retain their daytime colour. This is because light from the sun is white while that from sodium vapour in street lamps is just yellow. (There are other colours but they are so weak compared to the yellow component that they can be ignored.)

What is light?

We cannot understand colours in the sky unless we first understand about light.

Firstly, light is one narrow band of a very wide range of electromagnetic (EM) radiations. These are all alike in that they are waves but they differ because of their *wavelength*. You can use waves on water as a useful analogy to the invisible EM waves. It may seem strange to talk of light as invisible – but it is, until it reacts with something like the retinas of our eyes. You can see water waves going past but you cannot see light waves going past – they are, from that point of view, invisible.

The rate at which waves oscillate is their frequency – the number of times a second they vibrate. Wavelength multiplied by frequency is the speed at which the waves travel which, for EM waves, is always the same – 300 million metres per second. So waves of shorter wavelength are higher in frequency.

How energetic EM waves are depends on their wavelength. The shortest waves are of the highest energy and the most energetic with which we come into contact are gamma rays, like those that come from radioactive substances such as radium and from the sky in cosmic rays. Next come X-rays, then ultraviolet (UV) light, which our eyes cannot detect, followed by violet light, which they can. The UV is followed by the narrow band of wavelengths that (through our eyes) provide the sensations of colour and are the *visible spectrum* (Fig 17.1). Waves longer in wavelength than visible red make up the invisible infrared (IR) which contains the heat radiations.

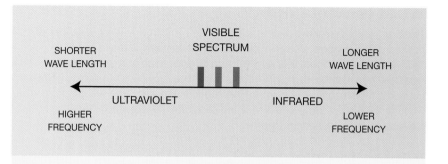

Fig 17.1 The visible spectrum is a very narrow band of wavelengths of light. Light is part of a broad spectrum of electromagnetic (EM) waves that extend from very energetic gamma rays to much less energetic radio waves. The latter bands do not have any direct effect on weather, whereas the ultraviolet and infrared (radiant heat) parts of the spectrum do.

Longer EM waves are used to broadcast TV, radio etc and are, except near their transmitters, weak (Fig 17.1).

Finally, in our understanding of colour in the sky we must come to grips with the packet of light energy called the *photon*. Changes of energy in atoms lead to the emission of photons, which we can visualise as small bundles of EM waves travelling away from the atoms at the speed of light. Each photon has its own wavelength which does not change, so we can talk of photons of red light, green light etc. Each photon can act like a snooker ball and collide with the electrons in atoms. It bounces off an atom but, unlike a snooker ball, during its collisions it can be absorbed by the atom and so disappear. These are illustrated in Fig 17.2.

White light consists of an immense number of photons of all the colours of the spectrum. On the other hand, the yellow light from sodium street lamps consists almost entirely of photons of yellow light. For this reason we call sodium vapour a *monochromatic* source of light.

Lamps of all kinds are emitting light but most of the colour we see about us is the result of white light falling on surfaces that absorb certain wavelengths and reflect others. When we see, say, a red rose, the rose is red because it is reflecting red photons (plus some other colours) and absorbing the photons of the rest of the spectrum. We can now begin to explain about colours in the sky – but the mechanism that gives the sky its blueness needs a different explanation.

What is white light?

The Sun sends us light of all the colours of the spectrum – what we often call the colours of the rainbow. The rainbow shows that sunlight is made of all the colours from red through orange, yellow, green and blue to indigo and violet. Sunlight, when reflected from raindrops, is split into its component colours but unless something happens (refraction in this case) to split the light like this, then we see sunlight as white and, by definition, *white light* is the sensation we experience when we see all the colours of the spectrum together.

Why is the sky blue?

Because the sky is always there, few people ask this question – but it is an interesting one. It obviously has to do with light from the sun but how is it that the sky is blue all over – even in the north? We can explain this and why the sky is blue – rather than any other colour – in Fig 17.2.

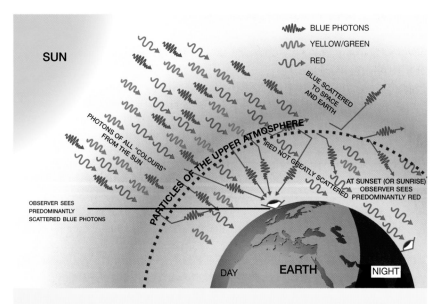

Fig 17.2 Why the sky is blue and sunset and dawn red. The sun sends us a full spectrum of all the colours of the rainbow. We have chosen three: red, green and blue. Particles in the high atmosphere are just the right size to intercept photons of blue light and scatter them. Thus we see these scattered blue photons in all directions and the sky is blue everywhere. However, if the evening and early morning skies are robbed of blue light this must leave red to dominate, giving red skies at night and in the morning.

We have picked out three representative colour photons from all those that are being emitted from the sun. As you go to very high altitudes, the atmosphere becomes very thin. Ordinary weather processes have long since been left behind.

All photons can make snooker-ball collisions with some of these rarefied air particles but those of certain wavelengths (colours) are much more likely to do so than others. In fact a blue photon is 16 times more likely to make a bouncing collision than a red one. These bouncing collisions are called *scattering*.

Thus, when the photons of light from the sun scatter off the particles of the high atmosphere, the blue ones are much more likely to get deflected sideways – even backwards – so it is scattered blue photons that come to you from all directions. On the other hand, the red ones (and other colours) tend to keep coming. Only if you look towards the sun – which you must not do – would you see the light that is robbed of much of its blue component.

Another consequence of this is that, seen from space, Earth is a blue planet because of the blue photons scattered back into space as well as those that are reflected from the Earth's surface, especially the sea.

Why are there red skies at night and in the morning?

This theory also explains why red is the predominant colour of dawn and sunset. As an example, take sunset. At sunset the sun is close to the horizon. Its light has had

Photo 55 A red sky in the morning is often a better weather predictor than one at night. There has to be a clear sky over the eastern horizon and the clouds have to be high. This occurs when a warm front or occlusion is advancing into the east, so the prediction of a bad day is often borne out.

Summer of '56

There was a prodigious storm in the summer of 1956 which reached its maximum intensity around the Isle of Wight. This is still remembered because at the same time there was a classic race for ocean-going yachts – the Channel Race – taking place in the area. There was also a lesser race for Firefly dinghies taking place from the yacht club on Hayling Island at the entrance to Chichester Harbour. I remember the evening before the big blow as I had hitched a ride home from Hayling, leaving my dinghy at the club. As we motored away from Hayling the western sky was literally a shade of purple, which I shall never forget. Experience has taught me that when the sky takes on strange colours then you can expect some very bad weather – in this case force 10 in Sea Area Wight.

to come through a wide slice of the atmosphere and so much of the blue light has been scattered to the west. This therefore leaves the light from the red end of the spectrum to come through.

At this time of day most of the action is either overhead or in the west. The same explanation goes for dawn, but now it is the clouds in the east that are predominantly illuminated with red light.

The clouds need to be fairly high at these times for the sun to shine on them and create the rosy-reds and other colours of vivid sunsets and dawns. Such high clouds are associated with fronts – cold fronts passing at evening or warm fronts approaching at dawn, or just islands and rafts of medium-level cloud about the sky that are probably the remains of old fronts.

Beautiful sunsets and dawns are not always red. More often they are shades of pink, yellow and orange (photo 8). The sunset is predominantly yellow when the air out to the west is clean and the more polluted it becomes –with smoke, fumes etc – the redder it becomes. Sunsets over the sea are often orange because of the effect of salt particles in the atmosphere. At dawn it is the cleanliness or otherwise of the air to the east that controls the predominant colours of the opalescent dawn sky.

Research done in the London area in the 1920s showed that a red dawn was followed by rain within 24 hours 70 per cent of the time and dry weather followed a red sunset with the same frequency.

How does a rainbow come about?

You can only see a rainbow with the sun behind you and sometimes, when the sun is high in the sky as it is around midday in summer, you may not be able to see one at all, depending on your latitude. The further north, the greater the chance of seeing rainbows at any time of day.

Secondary rainbow

When the sunlight is very intense, the light in the drops can go round again and come out as a secondary bow outside the primary one. It will be dim compared to the primary and its colours will now be reversed, with red on the inside. Only under the most favourable viewing conditions can a third bow be detected – I personally do not think I have ever seen one (Photo 56, page 134).

Obviously a rainbow says it is raining but not necessarily raining where you are. Rainbows are most likely to be seen in the afternoons and evenings of showery days. They form an arc in the sky which, if you were high enough up, would be a complete circle. As it is, rainbows have an end where they seem to disappear into the ground. This is where the proverbial crock of gold is to be found – but don't go looking for it because a rainbow is just an illusion. You move towards a rainbow and it moves away into the shower at the same rate because your own personal rainbow (everyone sees their own rainbow) is formed in the raindrops at a certain distance from your eye.

It is all to do with the *refraction* of light. Whenever light enters a transparent medium, be it glass or water, it becomes bent. To prove this, try the old 'stick in a pail of water' trick. Put a straight stick into a pail of water and it looks bent. You know very well it is not bent – this is because of the way water refracts (or bends) light.

Another thing that refraction does is to split light into its component colours. I always remember that *blue* is most *bent* by refraction. This means red is least bent.

A ray of sunlight enters a raindrop; it is white light and so composed of all the colours. The refraction bends the blue light most and the red least so, within the drop, it separates the colours. By a process we don't really need to go into, the different colour rays in the drop become totally reflected on the inside of the drop and, after another refraction, emerge to come back to you separated into their spectral colours. So we see blue on the inside and red on the outside of this primary bow.

Do rainbows tell us anything about the weather?

Well, according to sailors' weather lore they do:

> *Rainbow to windward*
> *Foul falls the day*
> *Rainbow to leeward*
> *Rain runs away.*

This means that foul weather follows when the wind in the morning is blowing towards the sun – a westerly situation. If showers are breaking out this early, it could

Photo 56 A double rainbow forms in a receding heavy shower. We see that red is on the outside for the primary bow and on the inside for the secondary bow.

well be a bad day. The second part of the jingle has the wind blowing away from the sun – still westerly, but now an evening situation. Again there is truth here. If the clouds are so broken in the west that the sun can form a rainbow in the showers that are disappearing into the east, then it presages a fair night.

Otherwise rainbows tell us that there are showers where the rainbow can be seen. In some cases when we have showers we may see parts of rainbows, usually near the ground and not very close. They tell us not to dismiss the showers yet.

Size matters

Not all rainbows exhibit the same colours – it all depends on how big the raindrops are. If the outer ring is a vivid red then the raindrops are about half a millimetre in diameter, ie they are quite large. If the drops are three times smaller then there is no red and the outer band is orange. If the droplets become very small then the bows that form in them tend to lose their colour. Fog has very small droplets and while it is perfectly possible to see a so-called fog-bow in a wall of approaching fog, this bow will appear colourless.

What is a sun dog?

It's important to realise that whenever we see rainbow colours somewhere in the sky, we are looking at sunlight coming through either plate-like ice crystals or water droplets. Often it is difficult, even for the experienced cloud observer, to tell whether high clouds are made of ice crystals (cirriform clouds) or water droplets (alto clouds). We can get a contribution from both when we see *sun dogs*.

Sun dogs are rainbow-coloured patches that appear either side of the sun (photo below) and at the same level. The fact that they are coloured shows at once that, just like the rainbow, they are due to refraction of light, but this time the refraction is through minute six-sided plates of ice.

Such plates form at lower altitude than the hexagonal prisms that give rise to ring haloes – thus we are often looking at layers where both ice and water exist together. We can therefore assume that the clouds are thin altostratus or altocumulus and so it seems likely that the higher clouds are increasing in depth, which indicates possible rain to come.

Photo 57 When there are thin alto clouds in the direction of the sun, we often see 'sun dogs' – iridescent patches either side of the sun and on its level, caused by refraction through the cloud droplets. It is quite rare to see both dogs at once as the clouds are often not the right density (here, the left one is much dimmer than the right). They usually mean the weather is about to change.

What's a halo?

There are haloes in the sky that one sees often and those that appear so rarely that you are lucky if you ever see them. They are almost always associated with ice-crystal clouds and show very little coloration. They can form circles, arcs and crosses in the sky, although such complex haloes are very rare. The most common, and the most useful for foretelling the weather, is the ring halo that forms about the sun or moon.

We can explain the ring halo as follows. The ice crystals that form cirrostratus clouds are most often of an elongated shape and they sink slowly in the sky. Thus, despite the fact that they orientate themselves randomly, they tend to fall with their long sides vertical. They are also hexagonal ice prisms and so transparent to sunlight, which refracts through them. Prisms transmit more light at certain angles than at others, and the angle for most cirrostratus prisms is such that it is seen at 22° to the sun. Because the crystals are spread in a veil across the sky, we see more light at 22° in all directions and so a more or less colourless ring appears encircling the sun. You can test if this is the common halo by spreading your fingers at arm's length. They should just fit between sun and halo (photo 15, page 41).

Harbinger of rain

Because cirrostratus is one of the clouds seen as a warm front (or occlusion) approaches, and fronts bring deteriorating weather, the ring halo is one of the best harbingers of coming rain. However, allow for the fact that rain does not always follow the sight of a halo. Someone may get rain but not necessarily you.

Photo 58 When the high atmosphere becomes laden with dust etc from a volcanic eruption, spectacular sunsets and dawns occur. This roseate glow, taken in Lubbock, Texas, was due to the eruption of Mount Pinatubo in the Philippines. Forest fires, on the other hand, lead to blue moons. Photo: Richard Petersen

Why is the moon blue?

After severe forest fires or big volcanic eruptions, the sky can exhibit very odd colours. Because of the effect of smoke and dust particles at high altitude, the normally blue sky can turn green or even red. If your sky does display unnatural colours, you will probably find the answer in the newspapers when they report big forest fires in some part of your latitude. For example forest fires in North America can lead to odd sky colours in Europe as the smoke particles are carried across the Atlantic by the upper winds.

The eruption of the volcano that destroyed the island of Krakatoa in 1883 is rated as the most violent in recorded history. Prodigious amounts of volcanic ash and dust were spewed into the atmosphere and at places, close to Krakatoa, completely blotted out the sun. The upper winds spread the fine dust around the world so that for years there were brilliant red sunsets as well as other unusual colours in the sky, such as green bands stretching up from the setting sun into the zenith.

The light from the moon will have its red component scattered by dust and smoke and so appear blue, but as this happens so rarely it gives rise to the saying:

Once in a blue moon.

Mount Pinatubo Volcano

This volcano in the Philippines came to life early in April 1991 after lying dormant for some 600 years. A series of eruptions blasted ash and smoke to 80,000 feet (15 miles or 24 km). This volcanic dust was transported around the world by the high altitude winds so that, for example, in Texas they began to see more red sunsets and dawns than usual in late July (photo 58).

Typically, 20 minutes after the sun had set, a great arch of dusty-red light filled the western sky extending to 30° above the horizon. A full three quarters of an hour after the sun had set, and while the rest of the sky continued to darken, this red region turned into a peach-pink oval at its highest point. After this display the colours gradually faded. Then, for the next couple of weeks the dust-induced coloured sunsets disappeared but came back again in August.

During the day the sun was sometimes surrounded by a glow whose radius was 30° and the sky looked as if it were covered in ripples. During August, weather satellites showed the northern edge of the dust cloud moving north from 25° to 30°N so places further north than this did not see the beautiful sunsets and dawns until later in the autumn. Similar effects appeared in central England in early December.

Whenever violent volcanic eruptions shoot dust and ash miles into the sky, places close to the same latitude can expect to see these kinds of unusual sky colorations.

18 what do trails tell us?

Those who are old enough to remember will tell you that the first time they became aware of the trails formed by high-flying aircraft was during the Battle of Britain. Ever since planes could operate at high altitude, they had been making trails – but only during the battles over southern England in 1940 did the sheer volume of the trails they made impress itself on everyone. The trails were not at all welcomed by those producing them because they made it quite obvious where the planes were. The planes themselves may have been specks in the sky but their exhaust trails ensured that friend and foe alike knew exactly where they were.

When the mighty Eighth Air Force bombers operated deep into hostile territory, it was pointless attempting to camouflage them because of the trails they left behind them – trails that, when the conditions were right, formed persistent areas of what in effect were cirrus clouds that outdid nature in their density.

When jet aircraft appeared, the trails became denser and more obvious and today if you live near any air corridor you will become fully aware of the high-flying passenger jets from the trails they produce. Otherwise the silvery specks would pass unnoticed more often than not (photo 59, page 139).

Why do planes produce trails?

Aircraft trails are often called *exhaust trails* because, with piston-engined planes, it is the engine exhausts that provide the extra moisture and the expansion required for the air to form cloud. Although we do not usually describe the efflux from jet engines as their exhaust, as the mechanism by which the vapour trails are made is the same as with piston engines so they can all be grouped under the head of exhaust trails – or simply trails.

The degree of wetness of the air through which the plane is flying will have a big effect on the trails themselves. If it is too dry, then no trail or only a very short one is formed. If very wet, then dense and persistent trails are formed. In between these two extremes the trails exist for a more or less short distance behind the aircraft before disappearing.

You can make cloud by doing two things – either create more vapour at a given temperature until the air can hold no more and deposits some of its excess as water droplets ie forms cloud. Otherwise you can make the air colder until it has to form cloud.

With exhaust trails you do both things. The combustion of petrol or aviation kerosene (as used in jets) creates moisture in the exhaust and the exhausted air expands, thus cooling down.

Can trails form at any height?

The answer is no. Below a certain height called the *Mintra height*, the air is too dry to allow trails to form.

Continue to ascend many thousands of feet and you reach a level where it is too cold (the *Maxtra height*) for trails to form. The zone between these two levels (which are always changing depending on the weather conditions) is where trails are made.

Photo 59 Two persistent trails cross one another. They show that the air at high-cloud level is saturated with water vapour and so we need to look around for the signs of invading cirrus. This can just be detected in the lower half of the photograph.

Trails and coming bad weather

The important message when cirrus cloud begins to invade the sky is that possibly there is deteriorating weather on the way. We saw in Chapter 5 that possibly the most important attribute of cirrus is the hours of forewarning it gives. Cirrus ahead of a coming depression can appear 12–24 hours ahead of when the deterioration will become a potential hazard.

Even before the first wisps of cirrus appear, dense trails will give you even longer warning that possibly there is some nasty weather in the offing. The trails are dense and persistent because the high-altitude air is relatively wet – something that is associated with coming warm fronts (photo 59).

The wind and the trails

We saw in Chapter 5 that when the weather is on the change the winds at altitude blow more or less across the wind direction near the surface. Specifically, the high-altitude wind comes from the left-hand side of the low altitude wind when deteriorating weather is on the way.

Photo 60 Once a trail is sown, what happens from then on can tell us a few things about the weather. Here a trail has shredded sideways, which shows that the wind up there is blowing across it, perpendicular to the wind that the weathervane indicates. When the upper and lower winds are crossed like this, the weather is on the change.

However, the reverse is also true. If the high altitude wind comes from the right of the low wind, then better weather is on the way. If the two wind directions are more or less parallel, do not expect much change in the situation.

Persistent trails help you to make up your mind about which way the wind aloft is blowing. With a trail you have something specific to watch and may well be able to see which way the trail is moving as a whole. However, there is another clue.

Watch the trails themselves. If they shred sideways, then the wind up there is across them. If they grow little turret-tops along their length, then the wind is along them (photo 60).

Knowing that the upper wind direction is one way or the other, it is then much simpler to make up your mind which way it actually is and you can make a forecast for better or for worse. Having done that, you cannot rest on your laurels. Keep an eye on the sky for better weather or worse weather clouds that will help to confirm or deny your forecast.

Trails at sunset and dawn

The trails that are made near sunset may help in predicting what sort of night it will be. If the aircraft have short – often very bright – trails then you can expect a good night to follow. If they sow long lengths of persistent trails across the sky, they may be pretty to watch but look to a night that will probably be blowing and raining by dawn.

Around dawn, aircraft that produce dense persistent trails are telling you that poorer weather is to be expected during the day but short trails that soon disappear presage a more or less fine day to follow.

19 hurricanes

Hurricanes do not occur in Europe but are endemic in the summer in the Caribbean area. This is because hurricanes are a product of warm tropical seas. Old hurricanes affect European shores in the late summer or early autumn when they re-curve back across the North Atlantic. By the time they get to Atlantic Europe, they are usually demoted to depressions but that does not prevent them from producing some typically tropical weather with very strong winds, heavy rain and thunderstorms.

What is a hurricane?

In Chapter 9 we read of the devastation caused in 1987 by the prodigious winds that swept the lands surrounding the Bay of Biscay and the English Channel. This became popularly known as the 'October hurricane'. But it was not a hurricane – the winds may have been hurricane force, but that did not make it a hurricane. The winds in the October hurricane did not rotate about a centre but just blew in more or less straight lines.

A hurricane is, in many ways, like a small, very intense, depression. However, the winds in depressions are prevented from becoming very strong because of the way the high-level winds blow over them. Winds up there usually blow west to east and may get as strong as 150–200 knots.

So, in a relatively short vertical distance, the circulating surface winds have to unwind into the upper westerlies. This puts a brake on how strong the surface winds can be – usually only as high as storm force. A hurricane is different. Where the hurricanes develop – towards the Equator in the North Atlantic – there are virtually no upper westerlies so, at all levels, the winds of hurricanes go round the same way. This enables the surface winds to become as strong as a hundred knots or more. Atlantic hurricanes usually start life as rather innocuous areas of low pressure somewhere near the Cape Verde islands and they develop as they move west. With time the wind picks up speed and rotates with increasing force around the centre, which is called the eye (photo 31, page 77).

The eye itself is cloudless but is surrounded by a vast ring of dense cloud. As the eye passes, the winds will go from being intensely strong from one direction to being just as strong from the reverse direction. This causes enormous seas and, together with the sudden change in wind direction, can cause vessels to founder.

How strong are hurricanes?

Hurricanes are categorised on a scale of 1 to 5 through the Saffir-Simpson Scale, which mainly depends on wind speed but also includes storm surges. They list the expected (or, after the event, the actual) effects as a hurricane makes landfall.

Category One Winds 64–82 knots (119–153 km/hr). Storm surge 4–5 ft above normal. Damage only to mobile homes and trees, some minor coastal flooding.

Category Two Winds 83–95 knots (154–177 km/hr). Storm surge 6–8 feet above normal. Roof etc damage to buildings. Some trees blown down. Some coastal areas flood 2–4 hours before eye of the hurricane arrives. Small craft break moorings.

Category Three 96–113 knots (178–209 km/hr). Surge 9–12 feet above normal. Damage to small buildings. Large trees blown down. Mobile homes destroyed. Low-lying coastal areas flood 3–5 hours before eye. Evacuation of residences near the shoreline may be required.

Category Four 114–135 knots (210–249 km/hr). Surge 13–18 feet above normal. Many roofs fail. All signs blown down. Complete destruction of mobile homes. Low-lying coastal areas flood 3–5 hours before eye. Major damage to lower floors of buildings near shore. Massive evacuation may be required up to 6 miles (10 km) inland.

Category Five Winds greater than 135 knots (249 km/hr). Surge generally greater than 18ft. Roofs of residences and industrial buildings fail. Some complete buildings demolished. All trees blown down. Complete destruction of mobile homes. Low-lying areas flood 3–5 hours before the eye. Major damage to lower floors of shore-side buildings which are less than 15ft above the sea level and within 500 yards (460m) of the shore. Massive evacuation within 5–10 miles of coast.

These latter storms are rare – only three have made landfall in the US since records began. Up to and including 2005 the most intense Category Five hurricane was Wilma, which had the lowest central pressure ever recorded.

Hurricanes onshore

Hurricanes feed on the warmth and moisture that they derive from the tropical waters they cross, but once ashore they rapidly lose their worst attributes. Their greatest danger to humanity is from the surges in sea levels they produce where the normal water level may rise by tens of feet. Then the sea drives well inland carrying boats, small ships, wooden buildings etc with it. The pounding waves add to the devastation.

Where do we find tropical storms?

Tropical storms are spawned over the warm seas of the tropics around the world. In the North Indian Ocean they are known as *severe cyclonic storms*. In the southwest Indian Ocean they are *tropical cyclones*. In the northwest Pacific west of the dateline they are *typhoons*. In the southeast Indian Ocean east of 90°E they are *severe tropical cyclones*. On the north coasts of Australia they are often called *willy-willies*. Only in the North Atlantic are they called *hurricanes*. They usually visit the Caribbean and the southeast coasts of the United States but members of a year's family of hurricanes come a little further north with time, so that the early ones affect the Caribbean and its islands while the later ones may come ashore on the Gulf Coasts of the United States. Occasionally they cross the isthmus of Tehuantepec and Yucatan to affect the west coasts of Mexico and up into California. Sometimes hurricanes will recurve and stay off the coast of the US when they may have already affected the islands etc of the Caribbean.

What happens to old hurricanes?

Not all hurricanes recurve into the Atlantic. Some move northwards across the eastern States, sometimes getting as far as Canada, or they may lose themselves inland somewhere.

The ones that recurve across the Atlantic get entrained by the upper-westerlies. The met services then demote them to depressions and are often reluctant to tell the public that the coming low is in fact an old hurricane. This is not good, because if you know that a depression coming out of the Atlantic in autumn is an old hurricane, you can allow for the weather it brings being more tropical in nature than you would expect. The air will sometimes be quite muggy, thunderstorms may break out when you did not expect them and rain is more showery and heavier than you thought it would be. The winds are often strangely gusty and stronger than anticipated.

These old hurricanes can get a new lease of life as they approach Atlantic Europe by drawing into their circulation some cold polar air, which rejuvenates them. They never become hurricanes again but they make for very lively depressions.

The sinking of the *Pamir*

On 21 September 1957 a tragedy occurred in mid-Atlantic due to an old hurricane. Five hundred miles west of the Azores the four-masted barque *Pamir* ran into old Hurricane Carrie. The *Pamir* was carrying grain from Buenos Aires to Hamburg but the loose grain shifted and the ship capsized with the loss of almost all of her crew, which included a majority of young sail-training cadets. The search for survivors by 50 ships lasted 9 days and involved 13 nations, but only 6 were rescued.

Allowing for hurricanes

If you live in the areas most at risk from hurricanes, and if you do not know what the latest advisories are and what to do when a big one threatens, then it is not for want of trying on the part of the National Oceanic and Atmospheric Administration (NOAA), who broadcast a torrent of public information when they detect that a hurricane they are following out in the Atlantic may endanger your area. Thousands of lives have been saved as the weather services have refined and broadened their techniques since the early days when the first weather satellites gave them the means to see where the hurricanes actually were and how they were moving.

The hurricane season is deemed to start in June and reaches a peak in September, to decline rapidly into the autumn. To allow for this those in the regions at risk are urged to have a cache of supplies such as food, torches (and batteries for them), etc. Yours may be an evacuation area, in which case, when advised by TV or radio, you will have to make a pile of your valuable possessions ready to load up the car and head inland. A Hurricane Watch issued for a part of the coast indicates that hurricane conditions are expected within 36 hours. (Check the Internet.)

Hurricane Katrina

The 2005 hurricane season was particularly vicious and the most active on record. There were 28 storms big enough to be given names, 15 of which were hurricanes (winds greater than 74mph) and of these, 6 struck the United States. Among these was Hurricane Katrina, the most deadly and costly hurricane, which hit the Bahamas, southern Florida, Cuba and the Gulf States, including New Orleans. Most of the eastern seaboard of the US felt its effects as the storm tracked northwards just off the coast. Approaching 2,000 lives were lost and the damage was estimated at over $81 billion. The flooding of New Orleans was a major disaster and led to much recrimination about the strength of the sea defences and the requisite authority's response.

20 tornadoes

Tornadoes are the most concentrated weather hazard on the planet. Hurricanes will be vastly larger and, all told, will potentially do more damage, create loss of life etc, but the effects of tornadoes are sometimes mind-numbing. What other phenomenon will elevate cars, buildings, trees etc and carry them, sometimes for miles? What other phenomenon can destroy whole townships, as happens on the Great Plains of the United States, leaving an area largely devoid of buildings? What can generate winds of such intensity that they can drive straws into the boles of trees as if they were thin metal nails? The only answer to these conundrums, and many others, is the tornado.

Photo 61 An unmissable sign that a thunderstorm could breed tornadoes is the sight of these downward-projecting spherical lumps. They are called 'mamma' because they resemble a cow's udder and may appear both on the leading edge and the trailing edge of the storm cloud.

Strength	Limit	Effects
F0 Weak	40–72 mph (64–116 km/hr)	Light damage
F1 Weak	73–112 mph (117–180 km/hr)	Moderate damage
F2 Strong	113–157 mph (181–253 km/hr)	Considerable damage
F3 Strong	158–206 mph (254–332 km/hr)	Extreme damage
F4 Violent	207–260 mph (333–416 km/hr)	Devastating damage

How do tornadoes form?

The parent of a tornado is a thunderstorm. Under certain circumstances what is called a *mesocyclone* forms within one of those big storms called *supercells*. The supercell is the strongest of storms but not all supercells produce tornadoes.

The mesocyclone is a tube of rapidly rotating air, which starts within the body of the parent supercell and, as it intensifies, emerges from the base of the cloud and descends towards the ground as the funnel of a tornado. It is when it touches down that a tornado begins to create damage (photo 62, page 149).

How strong are tornadoes?

Most European tornadoes form along the leading edges of cold fronts or under thunder clouds and are rated F0 or F1. Because, most often, they accompany the big gust of wind that appears at the head of an intense cold front, they have been dubbed 'gustnadoes'.

F0 AND F1

Tornadoes of strength F0 or F1 may blow your chimney pots down, take the roof off your garden shed and tear branches off trees in leaf. F1s may blow over weakly rooted trees. The strongest of them may even blow unprotected caravans over and damage thatched roofs. High-sided vehicles are at risk and cars may even be blown off the road. It has been found that most of these tornadoes only last a matter of 10 minutes or less before petering out.

F2

Occasionally in Europe, especially down towards Mediterranean latitudes, there are F2 tornadoes. Now large trees are blown down or snapped off, while campsites and caravan parks will be devastated. Mobile homes are very much at risk. The wind is sufficiently strong to turn small loose objects into missiles. In general these stronger tornadoes have a lifetime of some 20 minutes.

We have to allow for the trend, evident over the last decades, for European tornadoes to become more damaging than hitherto. As an example, the tornado that tore through a suburb of Birmingham in July 2006 was responsible for £40 million of damage and some 20 people were admitted to hospital. In the following month archaeologists sheltering in a heavy metal container in a quarry in Lincolnshire were injured when a tornado lifted and rolled the container. This tornado raised debris, including planks and pieces of metal, to as high as 200ft (61m). Even so, compared to what is routine in America this was a puny tornado, maybe only an F2.

F3

While very rare in Europe, the F3 tornado is well-known in 'Tornado Alley', that swathe of the Great Plains that sees the worst and most prevalent destructive tornadoes. Most houses will lose their roofs and even be demolished altogether. Most trees will be uprooted and the air is highly dangerous with flying debris. Cars can be overturned or even picked up bodily, while trains can be flipped over.

F4

Finally, the F4 'devastating' tornado sees whole villages and homesteads etc flattened. Livestock and humans can be elevated and carried considerable distances. The air is full of large flying objects and the only safe place is in a specially constructed cellar. Devastating tornadoes account for 70 per cent of all tornado deaths. The life of these tornadoes is typically greater than an hour.

F5

Professor Fujitsu included a category 5 tornado, which he described as 'incredible'. In this case, even steel-reinforced concrete structures are badly damaged and everything else is destroyed or carried away. The devastation may be such that it is impossible to decide whether this was an F5 or just a very bad F4. For this reason F4 and F5 tornadoes are being included under the same head in assessing their severity.

While the F5 tornado is so rare as not to register on the statistics, it is estimated that 29 per cent are F0; 40 per cent are F1; 24 per cent are F2; 6 per cent are F3 and less than 1 per cent are F4 in the United States. In Europe the percentage of F1/F2 tornadoes will together register close to 100 per cent as F3s are so rare.

The most destructive tornado ever reported

This occurred on 18 March 1925. It started at about 1pm in southeast Missouri and in about three hours travelled a straight and continuous path for nearly 220 miles (354 km) across southern Illinois to Indiana. During this time the surface cloud varied in width between a quarter and a whole mile and its speed of movement lay between 48 and 65 knots (24–33m/s).

In this most exceptional storm, the tornado cloud was so close to the ground that no one saw a funnel. The terrible toll included 695 dead and 2,027 injured. Three thousand houses were destroyed or damaged, including four small towns that were wiped out.

Photo 62 A big F3/F4 tornado approaching Roff in Oklahoma on 22 May 1984. Note the signs of debris in the air. Photo: courtesy of NOAA Photo Library.

When is the tornado season?

The season for tornadoes in the northern hemisphere is said to be from March to October and they are most likely to occur in the late afternoon and evening. Yet the tornado that damaged some 25 per cent of Selsey in West Sussex in 1998 occurred in January. Certainly in Tornado Alley the high season is from April to June but really there is no closed season for F0–F2 tornadoes. The season that sees the biggest supercell thunderstorms will be the one for strong or violent tornadoes; otherwise gustnadoes can occur with sharp cold fronts at any time of year.

Waterspouts

Over the sea, funnel clouds can be just truncated snakes of cloud hanging down from the overcast sky or they can be true tornadoes.

If they do not meet the sea surface then they just produce a whirlwind. However, over the sea, there is only water at the base of an active tornado and so the funnel becomes a whirling mass of water drops. The sea surface is disturbed and there is much spray. This is a *waterspout*.

Waterspouts may be innocuous or they may be very dangerous. In mid-Atlantic on 30 March 1923, the White Star liner *Pittsburgh* was struck by a huge waterspout that wrecked her bridge and filled her crow's nest with water. This spout contained a great mass of water 40–50ft (12–15m) wide and more than 70ft (21m) high. The damage was caused by the sudden collapse of the water in this spout onto the liner.

Yachts have sailed through spouts and their strange rotating winds to emerge wet but otherwise unscathed. Generally the waterspouts seen in European waters provide a spectacle but do no damage. However, strong ones have often been tornadoes before taking to the water, in which case they are called *tornado storm spouts*. The Selsey tornado (above) must have been spawned on the Isle of Wight and then reverted to a storm spout before becoming a tornado again on coming ashore at Selsey. Spouts may occur on large bodies of inland water but they are usually weak affairs.

A rare event in England

A damaging tornado is odd on the south coast of England. The coastal township of Selsey, south of Chichester in Sussex, had two within ten years.

The second, on the night of 7/8 January 1998, damaged 1,000 homes (20 per cent of Selsey), cutting a swathe a kilometre wide.

It came ashore from the direction of the Isle of Wight as a tornado storm-spout preceded by golf-ball-sized hail. The storm went on along the Sussex coast to damage further property.

Across the English Channel even stronger tornadoes affected the Calais area that same night. The previous tornado at Selsey occurred in November 1986 but was less intense and only damaged some 300 houses.

21 things worth knowing

Until now I have tried to avoid meteorological jargon as much as possible. However, it is impossible, in some cases, to describe the weather without knowing what the met-men mean by the terms they use, so here are some explanations and definitions that will help broaden the picture.

Advection When weather such as fog is blown by the wind from one place to another, it is said to be advected. (See Convection)

Air masses Depressions are formed by the clash of air masses. These are great blocks of warm air and cool air that tend to be warmer and cooler than each other throughout their depth, which may be 6 miles (10 km). An air mass involves thousands of cubic miles (or kilometres) of air with much the same characteristics. Cold air masses come originally from the **polar** regions and warm ones are called **tropical** as they stem from the subtropics. If they come mainly over water they are called **maritime**, while if they come mainly over land they are called **continental**. The coldest ones come directly from polar regions and are called **Arctic**. So air masses get names like **maritime polar**, the cool showery airstream that often comes from the north-west, or **continental tropical** which is the warmest and driest.

Anabatic winds These are winds that blow up hillsides and mountainsides that are in the Sun. (See Katabatic winds)

Anemometer A device for measuring wind speed.

Aneroid A form of barometer whose readings rely on the flexing of a slightly evacuated capsule as the air pressure changes.

Anticyclone or High A region of descending air and light winds, about whose centre the winds circulate in a clockwise direction. Highs will be mainly sunny in summer but are often depressingly cloudy in winter. Sometimes intense highs form across the normal path of the depressions and cut them off. Such highs are called **blocking highs**.

Aurora The great magnetic storms on the sun's surface, called 'sunspots', send charged particles across space and these collide with the rarefied atmosphere over the polar regions, resulting in light displays called the **Northern Lights** or the **aurora borealis**. They occur about 60 miles (100km) up and can be seen 600 miles (1,000km) away from where they are overhead. In Scandinavian countries they are sometimes called the **Storm Lights**, because they are associated with bad weather that follows within the next two weeks. During auroral displays, compass needles are deflected and short-wave radio transmissions are interrupted.

Barometer A device for measuring atmospheric pressure. Mercury barometers rely on the ability of the air pressure to support a column of mercury. As the pressure falls so the mercury does likewise. The legends of 'Fair', 'Change' etc on the faces of aneroid barometers are not very reliable and only the *tendency* of the change is important. Large downward (or upward) tendencies must result in increasing wind that may reach gale force. Otherwise, small wanderings of the barometric height often do not mean that there will be any great change in the weather pattern.

Convection Air that is warmed above its surroundings rises and cools. This is convection and heap clouds are often the result. Convection causes air currents in the vertical (both upward and downward) as opposed to advection which is movement in the horizontal.

Deepening A depression *deepens* when its central pressure falls.

Dew point This is the temperature to which air has to cool in order to form cloud, fog etc. The bases of cumulus clouds are where the air has fallen to the dew point. As its name suggests, it is also the temperature to which the ground must cool at night in order that dew can form.

Drizzle This is rain where the drops are less than half a millimetre in diameter. It falls from relatively warm clouds, unlike real rain, which has to start in regions where the temperature has fallen to low levels. Drizzle can occur in circumstances where rain is impossible. It is prevalent in hilly districts and is very saturating, as well as markedly cutting visibility.

Filling A depression is said to be filling when its central pressure is rising.

Heat islands Large towns and conurbations are generally warmer than the surrounding countryside because of the heat generated by dwellings, offices etc. When frost is likely to occur, often the towns remain above freezing.

Humidity Air always contains some water vapour. The more it contains, the higher its humidity. What is measured is Relative Humidity (RH), a measure of how wet the air is compared to it being saturated with water vapour. So an RH of 50 per cent means that the air contains just half the amount of vapour that would put it on the edge of forming fog. Humidity is measured by hygrometers. Simple ones can be bought from garden centres.

Instability   When the atmospheric conditions are such that air rises of its own accord, the air is said to be *unstable*. Unstable air leads to heap clouds, thunderstorms etc. The opposite of this is *stability*. Stability leads to layer clouds, continuous rain etc.

Jet streams   These are high-speed rivers of wind some 6 miles (10 km) high, where the wind speeds may be between 100 and 200 knots. They are associated with intense depressions which produce gales or severe gales at the surface.

Katabatic winds   These are winds that flow downhill overnight in hilly districts. In mountainous districts they may become quite strong. (See Anabatic winds)

Land breezes   These are winds that blow from land to sea overnight. Combined with katabatic winds they constitute *nocturnal winds*, which may be all the wind there is on some quiet nights.

Latent heat   It takes heat to turn water into vapour. The heat required to evaporate a litre of water when it is boiling at 100°C (212°F) is called its *specific latent heat*. If water vapour condenses back into water, this latent heat energy is given back again; so when clouds form they give out vast amounts of heat. This is one process by which inversion layers occur. Equally immense amounts of the sun's energy go to turning water from seas, lakes etc into vapour.

Radar   Weather radars are used to observe the position and intensity of rain etc. Most developed countries will have radars that cover their area and the results are to be found on the Internet.

Radiosondes   We know what is happening in the air above our heads because balloons bearing instruments are released across the world at the equivalent times to midday and midnight GMT. These measure temperature and pressure as well as humidity, and modern weather forecasting could not be done without their observations. They usually ascend to over 8 miles (13km), ie into the low stratosphere. They can be followed by radar and so give the speed and direction of upper winds. The latter are also monitored by balloons released at 0600 and 1800 GMT.

Satellites   Weather satellites are of two kinds, those that make circular orbits over the poles (polar satellites) and those that are placed at the correct distance above the Earth and over the Equator to remain more or less fixed over one spot (geostationary satellites). The cloud satellite pictures seen on TV are taken by the latter, while the former give meteorologists more information for their forecasts because they are much lower in the sky.

Seabreezes   Winds that blow from sea to land during the day. In most of Atlantic Europe they are, at most, moderate in strength but in Mediterranean latitudes they can get fresh or even strong.

Troughs   These are lines of deteriorated weather stretched across the wind. Frontal troughs are where the worst of the weather of fronts is concentrated but there are also air mass troughs, which normally form in cool airstreams and are lines of showers and associated cloud, again stretched across the wind. They are shown as plain thick lines on weather maps, with no protuberances.

Wave depressions   These are small lows that form on fronts, especially cold fronts and run along them much as a kink runs along a stretched rope when you give it a flick. They deteriorate the weather and shift the winds as they pass. Sometimes they develop into sizeable lows in their own right. Errors in the forecasts when the weather is poor are often due to the incidence of waves.

index